Systems Biology

Topical Conference at the 2009 AIChE Annual Meeting

Nashville, Tennessee, USA
8-13 November 2009

ISBN: 978-1-61567-919-5

Printed from e-media with permission by:

Curran Associates, Inc.
57 Morehouse Lane
Red Hook, NY 12571

CURRAN ASSOCIATES INC.
proceedings
.com

Some format issues inherent in the e-media version may also appear in this print version.

Printed by Curran Associates, Inc. (2010)

AIChE
3 Park Avenue
New York, NY 10016-5991

Phone: (203) 702-7660
Fax: (203) 775-5177

www.aiche.org

Additional copies of this publication are available from:

Curran Associates, Inc.
57 Morehouse Lane
Red Hook, NY 12571 USA
Phone: 845-758-0400
Fax: 845-758-2634
Email: curran@proceedings.com
Web: www.proceedings.com

TABLE OF CONTENTS

Antisense RNA, Transcriptional Interference and Genetic Switch: Train Wreck or Noise Filter?

Wei-Shou Hu, Chemical Engineering and Materials Science, University of Minnesota, Minneapolis, MN

Convergent overlapping genes with face to face promoters expressing cis sense-antisense transcripts (SAT) from the opposite strands of DNA are commonly seen in prokaryotic and eukaryotic genomes. The expression of these gene pairs is often reciprocally regulated through transcriptional interference (TI); a mechanism by which elongating RNA polymerases, like trains approaching from opposite directions, collide with each other. The wreckage caused by such collisions inevitably includes aborted RNA. Increasing evidence indicates that TI represents a unique mechanism in gene expression regulation. Many such systems are involved in controlling key biological events including the subjects of our study: prgX/prgQ gene pair in Enterococcus faecalis controlling the initiation of conjugation process transferring antibiotic resistance plasmid, and the scbA/scbR gene pair in Streptomyces coelicolor controlling the onset of antibiotic production. To understand this regulatory mechanism, a discrete mathematical model was developed to quantitatively assess the effect of RNA polymerase collision on transcription of the gene pair. Our model demonstrates that TI can give rise to a "switch-like" behavior, whereby depending on the state of the two promoters the system could switch between two distinct outcomes. The aborted transcripts produced by transcriptional interference in convergent promoter systems can potentially interact with opposing transcripts via sense/antisense interaction. Our data indicates that the aborted transcripts exist in various sizes depending on the loci of collision. Incorporating sense-antisense interaction between full length sense transcripts and aborted antisense transcripts in the model we demonstrate that aborted transcripts dampen the noise in such gene pairing systems, thus making such a switch behavior more robust. Given that sense/antisense pairs are conserved across species, TI is likely to have a significant regulatory role in gene expression. Our model suggests that they have the functionality to be gene switches.

Keywords: Transcriptional interference, Sense antisense transcripts, Antisense regulation, Convergent promoters, RNA polymerase collision, Mathematical model

See more of this Session: Paradigms in Systems Biology - Invited Session

See more of this Group/Topical: Topical A: Systems Biology

Stochastic Effects in Viral Gene Expression Contribute to HIV-1 Latency

David Schaffer, Department of Chemical Engineering, University of California at Berkeley, Berkeley, CA

HIV, the retrovirus that causes AIDS, can establish rare, latent infections of cells, and the resulting latent pools of virus represent the most significant barrier to elimination of virus from a patient since they persist for decades and can reactivate at any time. After HIV enters a cell, it semi-randomly integrates its genetic material into the host genome. The viral promoter then integrates several inputs to regulate the viral gene expression rate: epigenetic effects at the integration position, the level of activation of host signaling pathways and in particular host transcription factors that bind to the viral promoter, and the levels of virally-encoded proteins that feed back and regulate promoter activation. A high viral gene expression rate rapidly initiates viral replication, whereas low expression can lead to stochastic effects in viral gene expression that we have hypothesized contribute to viral latency.

We have conducted a rigorous experimental and computational analysis of the gene expression dynamics in a lentiviral model of HIV-1. This model system is derived from HIV-1 and is missing viral pathogenic functions but retains its ability to integrate chromosomally, its full enhancer and promoter elements, and its implementation of a positive feedback loop mediated by the virally-encoded factor Tat that strongly regulates transcriptional activation. We have previously reported that stochastic fluctuations in Tat lead to phenotypic diversity, or stochastic bimodality, in HIV-1 gene expression, a phenomenon that can be observed experimentally and predicted computationally with stochastic modeling. As a result, we have hypothesized that Tat-mediated feedback is activated sporadically such that after the virus infects a target activated T cell and integrates into the host genome, significant periods of time may elapse before viral expression reaches the point where reproduction and propagation can occur. This time could be enough to allow an activated T cell to transition to its memory state, thereby trapping the lentivirus in an inactive form until such time as that memory cell is reactivated.

While the majority of scientific effort is focused on the canonical variant of HIV that infects North America and Europe, however, given the rapid evolution of the virus in individual patients and across the world there is a staggering level of viral sequence diversity. We now build on our previous discovery with experimental and computational approaches that explore the question of how viral sequence diversity can impact viral transcriptional regulation and latency. These findings yield insights into the establishment of HIV-1 latency and may ultimately lead to the mechanistic understanding and design of therapies to purge and eradicate latent HIV-1 reservoirs.

See more of this Session: Paradigms in Systems Biology - Invited Session
See more of this Group/Topical: Topical A: Systems Biology

Robustness of Metabolic Networks. Some Fresh Insights

Hyun-Seob Song, School of Chemical Engineering, Purdue University, West Lafayette, IN

Doraiswami Ramkrishna, School of Chemical Engineering, Purdue University, West Lafayette, IN

We examine the analysis of robustness in the literature as applied to complex metabolic networks with regard to quantitative metrics and how they can influence metabolic engineering and associated areas of applications. The analysis is motivated by the extent to which results can be obtained without labor-intensive computation that is the hallmark of understanding complex systems.

See more of this Session: Paradigms in Systems Biology - Invited Session
See more of this Group/Topical: Topical A: Systems Biology

A Phenomenological Systems-Level Theory of Bacterial Cellular Metabolism

Luis A. N. Amaral, Chemical and Biological Engineering Department, Northwestern University, Evanston, IL

The metabolism of a number of organisms has been reconstructed from annotated genome sequences. Studies of constraint-based in silico models have shown that metabolic states are the result of the optimization of a biological function, generally growth, acquired through evolution. The in silico models are also able to make accurate predictions of cellular growth. Despite these advances, we still lack a systems-level mathematical theory that is able to predict bacterial growth in arbitrary growing media and that is easily generalizable.

Here,we develop a phenomenological theory that describes the relationship between input (nutrients) and output (growth) for bacterial metabolism and test it against in silico predictions for Escherichia coli. Our theory suggests that, to maximize growth, the metabolism of E. coli effectively distributes nutrient uptake fluxes equitably among pathways that result in synergistic interactions with other nutrients.

Additionally, our theory provides a natural classification of nutrients into groups with different impacts on growth and opens the door to predicting the growth rate of any microbe with the need for only a few experiments.

See more of this Session: Paradigms in Systems Biology - Invited Session
See more of this Group/Topical: Topical A: Systems Biology

Designing Effective Trials Regarding Treatments against Cancerous TUMOR Growth

Ileana C. Carpen, Department of Chemical Engineering, Tennessee Tech University, Cookeville, TN

Nemoy Rau, Department of Chemical Engineering, Tennessee Tech University, Cookeville, TN

Cellular adaptation to changing environments is intricately coordinated through many molecular and mechanical responses. As cells evolve, abnormal growth patterns occur (that cannot be controlled by normal mechanisms) leading to cancer. With mathematical modeling we can integrate different characteristics of tumor growth for a non-experimental study of cancer.[1] Through integration of some of the prior clinical, experimental, and mathematical studies, we utilize a cellular automata model to take into account multiple factors affecting tumor growth in healthy tissue.[2] This in silico simulation model of tumor growth is based upon molecular and life cycle features that effect the growth rates of cancer. The life cycle parameters used in this model include replication rate, nutrient concentrations for proliferation and survival, and possible drug effects.[3] This 3-D in silico model is used to study the effects of drugs upon the growth and development of the tumor and optimize a drug treatment protocol under various conditions.

References

1. Quaranta V, Cummings P, Anderson A. (2005). "Mathematical modeling of cancer: The future prognosis and treatment." Clinical Chimica Acta 357,173-179.

2. An-Shen Q, Xiang Z, Chan Y, Bao-Sheng A. (1993). "A Cellular Automaton Model of Cancerous Growth." J. Theor. Biol. 161, 1-12.

3. S.C. Ferreira, Jr., M.L. Martins, M.J. Vilela (2002). "Reaction-diffusion model for the growth of avascular tumor." Phys. Rev. E65, 021907.

See more of this Session: In Silico Systems Biology: Cellular and Organismal Models
See more of this Group/Topical: Topical A: Systems Biology

Agent-Based Simulation of Endotoxin Induced Acute Inflammatory Response in Human Blood Leukocytes

X. Dong, Biomedical Engineering, Rutgers University, Piscataway, NJ

P.T. Foteinou, Biomedical Engineering, Rutgers University, Piscataway, NJ

S.E. Calvano, Department of Surgery, UMDNJ-Robert Wood Johnson Medical School, New Brunswick, NJ

S.F. Lowry, Surgery, UMDNJ-Robert Wood Johnson Medical School, New Brunswick, NJ

Ioannis P. Androulakis, Biomedical Engineering, Chemical & Biochemical Engineering, Rutgers University, Piscataway, NJ

The acute inflammatory response is one important component of the initial response of the host to a diverse array of biological stressors including infection, burns, trauma and surgery. While the amount of experimental information has grown, the complexity of the response has hampered the search for adequate clinical approaches in the settings of various inflammation related disorders. A key reason for this conundrum is the difficulty of predicting the impact of manipulating individual components of the highly complex, non-linear and redundant inflammatory response[1]. Progress in treating these processes requires a greater understanding of how components are organized to generate a behavior thus making model based approaches appealing[2]. As a result, various approaches have been proposed to simulate the underlying complexity of the inflammatory response including both equation based models (EBM) and agent based models (ABM)[3].

Although both modeling approaches (ODE and ABM) might provide significant insight regarding the behavior of complex biological processes, the underlying complexity of the dynamics of AIR pose a significant challenge in that the conventional reductionist approach using continuous, deterministic equation based models (EBM) is insufficient to capture the stochastic nature and dynamic transitional states in biological systems[4]. The recognition that nature is predicated upon cellular processes with stochastic properties has benefited the development of agent based modeling. Agent based models reflect the discrete stochastic nature of interactions and provides a realistic description of sub-cellular events. The key elements in ABMs are the agents, which are entities that represent a certain aspect of the system, for instance a family of cells and/or molecules, who are able to adapt and interact with the environment and with each other based on a specific set of rules[5].

To meet this challenge, an agent based modeling framework (ABM) is proposed to study the complex, non-linear dynamics of human acute inflammation. Predicated upon the essential interactions that define the development of deterministic human inflammation models[6, 7], we opt to translate them into an integrated ABM framework. Accordingly, interacting agents involve either inflammation specific molecules or cells essential for the propagation of the inflammatory reaction across the system. Therefore, spatial orientation of molecule interactions involved in elaborate signaling cascades as well as cellular heterogeneity are particularly taken into account. The proposed ABM framework is naturally stochastic in that interactions are designed to be based upon probabilities providing significant insight on how the stochastic fluctuations of molecular species aggregate to engender the overall systemic response.

The proposed in silico model is evaluated through its ability to reproduce a self-limited inflammatory response successfully as well as a series of scenarios indicative of the non-linear dynamics of the response. These scenarios involve either a persistent (non)infectious response or tolerance and potentiation effects followed by perturbations in intracellular signaling molecules and cascades. Such modeling effort would potentially provide significant insight on the stochastic interactions of the species involved in the propagation of the inflammatory reaction on the cellular response level; thus making it a critical enabler for improving our understanding of how manipulating the behavior of molecular species could manifest into the emergent behavior of the overall system.

1. Vodovotz, Y., et al., Mechanistic simulations of inflammation: Current state and future prospects. Math Biosci, 2008.

2. Clermont, G., et al., In silico design of clinical trials: a method coming of age. Crit Care Med, 2004. 32(10): p. 2061-70.

3. Vodovotz, Y., et al., Mathematical models of the acute inflammatory response. Curr Opin Crit Care, 2004. 10(5): p. 383-90.

4. Lowry, S.F., et al., Challenges for modeling and interpreting the complex biology of severe injury and inflammation. J Leukoc Biol, 2008. 83(3): p. 553-7.

5. Vodovotz, Y., et al., Translational systems biology of inflammation. PLoS Comput Biol, 2008. 4(4): p. e1000014.

6. Foteinou, P.T., et al., Modeling endotoxin-induced systemic inflammation using an indirect response approach. Math Biosci, 2008. 217(1): p. 27-42.

7. Foteinou, P.T., et al., In silico simulation of corticosteroids effect on an NFkB- dependent physicochemical model of systemic inflammation PLoS ONE, doi: 10.1371/journal.pone.0004706 2009 4(3).

See more of this Session: In Silico Systems Biology: Cellular and Organismal Models
See more of this Group/Topical: Topical A: Systems Biology

Molecular Level in Silico Analysis of Mass and Energy Flows in Microbial Communities

Ross P. Carlson, Chemical and Biological Engineering, Montana State University, Bozeman, MT

Reed Taffs, Chemical Engineering, Montana State University, Bozeman, MT

Complex microbial communities drive the Earth's biogeochemical cycles. In spite of their importance, the biochemical interactions within these communities are not yet well understood, nor are many in silico methodologies available for studying them. Three in silico methodologies based on stoichiometric network analysis were developed for studying mass and energy flows in microbial communities on the molecular level. Each approach has distinct advantages and disadvantages suitable for analyzing systems with different degrees of complexity and different levels a priori knowledge. These methodologies were tested and compared using the extensive data from the phototrophic, thermophilic mat communities at Octopus and Mushroom Springs in Yellowstone National Park. The models included three community guild members: cyanobacteria, filamentous anoxygenic phototrophs, and sulfate reducing bacteria. The in silico models were used to explore fundamental microbial ecology questions including the prediction and explanation for measured relative abundances of the oxygenic phototrophic primary producer cyanobacteria and the filamentous anoxygenic phototrophic bacteria. The three approaches represent a flexible toolbox which can be rapidly adapted to study other microbial systems with a variety of electron donors and acceptors on scales ranging from individual cells in a pure culture to entire ecosystems represented by metagenomic data.

See more of this Session: In Silico Systems Biology: Cellular and Organismal Models
See more of this Group/Topical: Topical A: Systems Biology

Metabolic Reconstruction and Analysis of Butanol Producing Strain Clostridium Beijerinckii NCIMB 8052

Caroline B. Milne, Chemical and Biomolecular Engineering, University of Illinois Urbana-Champaign, Urbana, IL

Ravali Raju, Chemical and Biomolecular Engineering, University of Illinois Urbana-Champaign, Urbana, IL

Pan-Jun Kim, Institute for Genomic Biology, University of Illinois Urbana Champaign, Urbana, IL

Nathan D. Price, Department of Chemical and Biomolecular Engineering, Institute for Genomic Biology, University of Illinois Urbana Champaign, Urbana, IL

Butanol is a higher carbon alcohol with lower water affinity and vapor pressure than ethanol, and hence holds great promise as a liquid fuel alternative. Clostridium beijerinckii has the potential to make butanol economically competitive because it naturally produces butanol as a byproduct of its metabolism and co-ferments pentoses and hexoses (primary components of lignocellulosic hydrolysate). We reconstructed the metabolic network of C. beijerinckii NCIMB 8052 using a combination of novel automation tools and publically available network curation algorithms (GapFill, GapFind). This semi-automated approach to building a computable genome scale model iteratively makes use of published in silico models to supplement network information from more traditional genome annotation databases. The reconstructed network comprises 641 enzymatic genes that catalyze 703 reactions and is further supported by metabolomics measurement data and literature evidence. Constraint-based analysis on the metabolic model reveals potential strategies to rationally align the natural objectives of the organism (selection pressure for growth) with the industrial objectives of the engineer (butanol production). We performed simulated gene knockout experiments to optimize C. beijerinckii's stoichiometric ability to produce butanol from minimal media. The model therefore serves as a guide to experimentally engineer network states in C. beijerinckii to produce an economically competitive fermentation organism.

See more of this Session: In Silico Systems Biology: Cellular and Organismal Models
See more of this Group/Topical: Topical A: Systems Biology

Constraint-Based Analysis of Metabolic Capacity of Salmonella Typhimurium LT2

Jennifer L. Reed, Department of Chemical and Biological Engineering, University of Wisconsin-Madison, Madison, WI

Anu Raghunathan, Division of Infectious Diseases, Mt Sinai School of Medicine, New York, NY

Sookil Shin, Division of Infectious Diseases, Mt Sinai School of Medicine, New York, NY

Bernhard O. Palsson, University of California, San Diego, La Jolla, CA

Simon Daefler, Division of Infectious Diseases, Mt Sinai School of Medicine, New York, NY

Computational models of microorganisms be useful for integrating and analyzing various types of data, and for making predictions about cellular behavior in different genetic or environmental backgrounds. We have developed a genome-scale metabolic reconstruction for *Salmonella typhimurium* LT2, a human pathogen that is closely related to *Escherichia coli*. The reconstructed metabolic network for *S. typhimurium* contains 1,087 metabolic and transport reactions connecting 759 metabolites. The resulting network accounts for nearly a quarter of the genes in the bacterium (1,083 out of 4,489). We have translated this reconstruction (iRR1083) into a constraint-based metabolic model for *S. typhimurium* LT2. We used this model to analyze experimental data taken from minimal media conditions, including growth rates, growth conditions, and gene essentiality. Comparisons between *S. typhimurium* and *E. coli* model predictions provides examples where these two closely related organisms differ in their metabolic capabilities and regulation of enzyme activity. In addition to minimal media conditions, we have also simulated growth in rich media (LB or a host-cell environment) to identify genes needed for virulence and to investigate which metabolic reactions could be utilized under such conditions and compared this with gene expression and proteomic measurements. We will discuss how genomic information can be used to develop models that can be used for predicting phenotypes, analyzing experimental data, and generating hypotheses about metabolism and regulation.

See more of this Session: In Silico Systems Biology: Cellular and Organismal Models
See more of this Group/Topical: Topical A: Systems Biology

Genome Scale Modeling of Pseudomonas Aeruginosa in a Microbial Air-Cathode Single Chamber Fuel Cell

Andrés Fernando González Barrios, Grupo de Diseño de Productos y Procesos (GDPP), Department of Chemical Engineering, Universidad de los Andes, Bogota, Colombia

Juan Diego Mejía Méndez, Grupo de Diseño de Productos y Procesos (GDPP), Department of Chemical Engineering, Universidad de los Andes, Bogota, Colombia

Felipe Racines Pérez, Grupo de Diseño de Productos y Procesos (GDPP), Department of Chemical Engineering, Universidad de los Andes, Bogota, Colombia

Nubia Milena Velasco Rodríguez, Centro de Optimización y Probabilidad Aplicada (COPA), Department of industrial Engineering, Universidad de los Andes, Bogota, Colombia

María Teresa Cortes Montañez, Laboratorio de Electroquímica y Materiales Poliméricos, Department of Chemistry, Universidad de los Andes, Bogota, Colombia

Martha Josefina Vives-Flórez, Centro de Investigaciones Microbiológicas (CIMIC),Department of Biological Sciences, Universidad de los Andes, Bogota, Colombia

Microbial fuel cells (MFC) represent nowadays a significant alternative energy source for low power demand devices such as medical implants or biosensors. Its advantage is based on the fact that it is possible to obtain energy from organic substrates at low cost. Nevertheless, there exist several technological limitations that do not permit to expand its applications such as the development of efficient electrodes or the selection of suitable microorganisms. In order to obtain a better understanding of the performance and the effects of the metabolism on the cell we propose a 2D air-cathode single chamber microbial fuel cell model in COMSOL®. This model accounts for the effects of metabolites diffusion in the chamber. In order to consider for the metabolism state of the cell we coupled this model with a genomic scale model of *Pseudomonas aeruginosa* based on flux balance analysis (FBA) as a lineal programming model implemented in Xpress MP. We built an objective function composed by the biomass flux of the cell and the electron shuttle flux in order to make both fluxes positive; by maximizing this function we could predict time profiles of the power density and these profiles were then validated in a 30 ml fuel cell with glucose as a substrate. Finally, *in silico* knockouts were carried out to determine genes deletion candidates to increase the electron shuttles such as pyocianin and phenazine. The *in silico* model predicted maximum potentials of 0.135V that were consistent to the ones obtained in the experimental cell.

See more of this Session: In Silico Systems Biology: Cellular and Organismal Models
See more of this Group/Topical: Topical A: Systems Biology

Deterministic and Stochastic Population Level Simulations of An Artificial *Lac* Operon Genetic Network

Michail Stamatakis, Department of Chemical and Biomolecular Engineering, Rice University, Houston, TX

Kyriacos Zygourakis, Department of Chemical and Biomolecular Engineering, Rice University, Houston, TX

The *lac* operon genetic switch is considered as a paradigm of genetic regulation. This system has a positive feedback loop due to the LacY permease boosting its own production by the facilitated transport of inducer into the cell and the subsequent de-repression of the *lac* operon genes. We present computational studies of an artificial *lac* operon network that highlights this positive feedback mechanism. Our single cell model takes into account the constitutive expression of the *lacI* repressor gene, the expression of LacY from an IPTG inducible promoter and the free and facilitated transport of IPTG through the cell membrane. This single cell model is incorporated into two Monte Carlo simulation frameworks that describe cell population dynamics. The first framework assumes deterministic reaction dynamics and takes into account stochastic DNA duplication, division and partitioning. The second framework takes into account stochastic reaction occurrence in addition to the aforementioned sources of stochasticity. By comparing the results of these two approaches, we elucidate the effect of genetic network structure and that of stochasticity in shaping the phenotypic distributions. We further investigate whether the salient features of these distributions can be captured with simpler single-cell models.

See more of this Session: In Silico Systems Biology: Intracellular Signaling and Gene Regulation
See more of this Group/Topical: Topical A: Systems Biology

Stochastic Simulation of Cytosolic Calcium Dynamics

TaiJung Choi, Department of Mechanical and Aerospace Engineering, University of California, San Diego, La Jolla, CA

Mano R. Maurya, Department of Bioengineering, University of California San Diego, La Jolla, CA

Daniel M. Tartakovsky, Department of Mechanical and Aerospace Engineering, University of California, San Diego, La Jolla, CA

Shankar Subramaniam, San Diego Supercomputer Center, Department of Bioengineering, Department of Chemistry & Biochemistry, University of California San Diego, La Jolla, CA

Ordinary differential equations-based deterministic simulation is accurate only if large reaction volumes (and hence a large number of molecules) are considered. At the subcellular level, the stochastic behavior of chemical reactions becomes effective. This is even more important when dealing with local (spatial) variations. Many of the cellular functions are strongly dependent on the spatial variation of chemical species inside the cell. For example, the density of receptors on the cell surface of a cell is not uniform. Instead, patches of high-receptor density are found. So, to account for spatial variations, in the deterministic approach, the system is modeled as a set of partial differential equations. The resulting equations are solved using the grid-based finite elements method. The number of species molecules in each of the grids is even smaller. Thus, stochastic simulation is necessary.

Given that any of the realizations of stochastic simulation do not individually correspond to the experimental results in single cells, at best, the average and other statistical measures, such as distinct clusters if any, of many (thousands of) realizations can be compared to the corresponding properties of a small cell-population. This results in a more accurate description of the system as compared to that provided by deterministic simulation. However, the computational complexity is enormous, thus requiring the methods for accelerated stochastic simulation. Starting with the pioneering work of Gillespie (1) and several other researchers, recently several methods have been developed to tackle the complexity of stochastic simulation. One approach is the tau-leap method in which a larger time-step is taken and many reactions are fired in the chosen time interval (2). A variant of this method is the binomial tau-leap method to avoid negative populations (3, 4). In another approach, deterministic simulation, stochastic simulation and Markov model based approaches are used in an integrated manner (5). Yet a third approach partitions the reactions and species as "slow" and "fast" with the partial equilibrium assumption that the population of slow species is not altered by the fast reactions (6).

Here we present the application of stochastic simulation technique to a model of the regulation of the dynamics of cytosolic calcium ions (Ca2+) in mouse macrophage RAW 264.7 cells that we have developed recently (7). The (current) deterministic model consists of 19 ordinary differential equations. The flux expressions are based on simple rate laws such as law of mass action (LOMA) kinetics, Michaelis-Menten (M-M) kinetics and Hill-dynamics as well as more complex forms. We present the results of stochastic simulation and compare the response characteristics obtained in stochastic and deterministic simulation.

Literature Cited

1. Gillespie, D. T. 1976. General method for numerically simulating stochastic time evolution of coupled chemical-reactions. Journal of Computational Physics. 22:403-434.

2. Gillespie, D. T. 2001. Approximate accelerated stochastic simulation of chemically reacting systems. Journal of Chemical Physics. 115:1716-1733.

3. Chatterjee, A., D. G. Vlachos, and M. A. Katsoulakis. 2005. Binomial distribution based tau-leap

accelerated stochastic simulation. J Chem Phys. 122:024112.1-7.

4. Chatterjee, A., and D. G. Vlachos. 2006. Multiscale spatial Monte Carlo simulations: Multigriding, computational singular perturbation, and hierarchical stochastic closures. J Chem Phys. 124:064110.1-16.

5. Kaznessis, Y. N. 2006. Multi-scale models for gene network engineering. Chemical Engineering Science. 61:940-953.

6. Cao, Y., D. Gillespie, and L. Petzold. 2005. Multiscale stochastic simulation algorithm with stochastic partial equilibrium assumption for chemically reacting systems. Journal of Computational Physics. 206:395-411.

7. Maurya, M. R., and S. Subramaniam. 2007. A kinetic model for calcium dynamics in RAW 264.7 cells: 1. Mechanisms, parameters, and subpopulational variability. Biophys J. 93:709-28.

See more of this Session: In Silico Systems Biology: Intracellular Signaling and Gene Regulation
See more of this Group/Topical: Topical A: Systems Biology

Quantitative Analysis of the Gene Regulatory Network Activated by Angiotensin II Type 1 Receptor in the Brainstem

Gregory M. Miller, Department of Chemical Engineering, University of Delaware, Newark, DE

Rajanikanth Vadigepalli, Daniel Baugh Institute for Functional Genomics and Computational Biology, Thomas Jefferson University, Philadelphia, PA

James S. Schwaber, Daniel Baugh Institute for Functional Genomics and Computational Biology, Thomas Jefferson University, Philadelphia, PA

Babatunde A. Ogunnaike, Department of Chemical Engineering, University of Delaware, Newark, DE

The angiotensin II type 1 receptor (AT1R) in the brainstem mediates a wide range of autonomic functions, and the dysregulation of AT1R signaling causes serious diseases such as hypertension. However, the molecular mechanisms underlying AT1R function in the brain are not well- understood, and an improvement in understanding is necessary in order to develop effective treatments of hypertension. In this work, we investigate the dynamics of the AT1R activation of tyrosine hydroxylase-a critical mechanism for the central regulation of blood pressure-by taking a quantitative modeling and analysis approach.

Using the regulatory mechanisms established in the literature (Veerasingham and Raizada, 2003), we constructed a mechanistic, ordinary differential equations model for the induction of tyrosine hydroxylase gene expression modulated by the gene regulatory network activated by AT1R. Our model simulations predict that AT1R regulates tyrosine hydroxylase gene expression by activating a combination of AP-1 transcription factor dimers. Specifically, c-Fos:c-Jun is predicted to be the dominant AP-1 dimer activated during the first 40 minutes of AT1R stimulation, with c-Jun:c-Jun subsequently becoming the dominant AP-1 dimer in the long-term AP-1 response. By predicting an activation of multiple AP-1 dimers, our simulation results suggest that the AT1R induction of tyrosine hydroxylase utilizes redundant mechanisms to ensure robustness in the face of perturbations in network components. To understand how reaction mechanisms influence model output combinatorially, we performed a multivariate sensitivity analysis of the AT1R gene regulatory network (Saltelli, 2002; Hua et al., 2006). This model analysis suggests that the initial activation of AP-1 is sensitive to reactions related to c-Fos activation and c-Fos:c-Jun dephosphorylation, while the long-term activation of AP-1 is sensitive to mechanisms related to c-Jun activation. These predicted shifts in sensitive network nodes are supported by experimental data from pathway inhibitor studies. The implications of these sensitivity analysis results are that manipulating the AT1R induction of tyrosine hydroxylase gene expression would require affecting specific reaction modules at particular times. In our presentation, we will discuss the model development, present a detailed analysis, and suggest alterations in specific reaction mechanisms to manipulate AT1R network function. For instance, we will present our predictions describing how perturbations leading to increased c-Jun activation without increased FRK signaling would lead to a shift of the activated AP-1 dimers to favor c-Jun:c-Jun. Such quantitative insights into the dynamics of key molecular mechanisms underlying blood pressure control mechanisms as these could provide new targets for developing effective treatments of hypertension.

References

Hua F., Hautaniemi S., Yokoo, R., Lauffenburger D.A (2006) **Integrated mechanistic and data-driven modeling for multivariate analysis of signaling pathways.** *J. R. Soc. Interface.* **3**(9): 515:26.

Saltelli, A. (2002) **Making best use of model evaluations to compute sensitivity indices.** *Comp. Phys. Comm.* **145**, 280-297.

Veerasingham S.J. & Raizada M.K. (2003) **Brain renin-angiotensin system dysfunction in hypertension: recent advances and perspectives.** *Br. J. Pharmacol.* **139**(2):191-202.

See more of this Session: In Silico Systems Biology: Intracellular Signaling and Gene Regulation
See more of this Group/Topical: Topical A: Systems Biology

Using Mutual Information to Identify TAR-Tat Coevolution in HIV-1 and Its Implications for Viral Latency

Siddharth S. Dey, Department of Chemical Engineering, University of California at Berkeley, Berkeley, CA

John C. Burnett, Department of Chemical Engineering, University of California at Berkeley, Berkeley, CA

Adam P. Arkin, Bioengineering, University of California at Berkeley, Berkeley, CA

David V. Schaffer, Department of Chemical Engineering, University of California at Berkeley, Berkeley, CA

After Human Immunodeficiency Virus (HIV-1) integrates into the host chromosome, the provirus displays two distinct gene expression states - a transactivated state resulting in the formation of new viral particles or an inactive state resulting in the establishment of latency. The position of integration, the recruitment of activating and repressing factors by the viral promoter, and the viral protein Transactivator of Transcription (Tat) have been shown to be the most important factors in determining the gene expression state of the provirus. The few Tat molecules that are synthesized from weak viral gene expression shortly after integration bind to a stem-bulge-loop RNA structure called Transactivation Response Element (TAR) and recruit other cellular factors to promote the phosphorylation of RNA polymerase II and result in transcriptional activation, whereas the failure of these few Tat molecules to initiate the positive-feedback loop results in transcriptional silence and latency. The stochastic nature of this positive feedback loop thus gives rise to these two distinct gene expression modes with dynamics switching between them. Numerous studies that have examined the impact of the positive-feedback loop on viral gene expression and latency have primarily been limited to subtype B, the most prevalent subtype in the United States and Western Europe. In this paper, we have applied computational and experimental approaches to analyze the impact of TAR and Tat sequences from six different global isolates (A, A2, B, B/F, C and D) and the coevolution between different sites on viral gene expression and latency.

We constructed an open-loop system in which gene expression from the HIV-1 Long Terminal Repeat (LTR) with a particular TAR subtype was measured using Green Fluorescent protein (GFP) (called the LG vector(1)), and Tat was placed under the control of an ubiquitin promoter and its expression level quantified using the fluorescent protein mCherry (called the Ub-ChIT vector). Plots of gene expression vs. Tat concentration for certain TAR-Tat pairs were sigmoidal in nature, suggesting that large concentrations of these Tats are required to upregulate gene expression and that these Tats could potentially give rise to large populations of latent reservoirs. In contrast, some other pairs showed high levels of gene expression at even low concentrations of Tat. The impact of different TAR-Tat pairs on gene expression and latency was then studied in the context of the closed-loop system using the LTR-GFP-IRES-Tat (LGIT) lentiviral system(1). In agreement with the open-loop system, weaker TAR-Tat pairs had a significantly larger number of latent cells as compared to the stronger TAR-Tat pairs.

To understand the impact of sequence variation on expression dynamics and latency, we employed a statistical method called mutual information to identify positions that may be coevolving in TAR and Tat. An analysis of coevolving residues within Tat revealed that certain amino acids positions were strongly correlated. To test the in silico results, mutations introduced at these positions to give an evolutionarily non-conserved sequence resulted in a very weak Tat, with the majority of the provirus in an inactive gene expression state. Finally, a search for positions that are coevolving within TAR and Tat revealed that nucleotides in the bulge of TAR are coevolving with amino acids in the Arginine Rich Motif (ARM) of Tat.

Thus we have shown that TAR and Tat sequences from different subtypes display a wide array of gene expression levels, with certain TAR-Tat pairs having a very high propensity for latency. Furthermore, statistical analyses using mutual information has revealed that certain positions within TAR-Tat and in Tat alone are strongly correlated and coevolution within these positions are responsible for these unique gene expression phenotypes, yielding fundamental insights into mechanisms of viral evolution and behavior.

References:

(1) Weinberger, L. S., Burnett, J. C., Toettcher, J. E., Arkin, A. P. & Schaffer, D. V. (2005) Cell 122, 169-82.

See more of this Session: In Silico Systems Biology: Intracellular Signaling and Gene Regulation
See more of this Group/Topical: Topical A: Systems Biology

Endocytosis and Signaling: Insilico Modeling at Multiple Scales

Neeraj Agrawal, Chemical and Biomolecular Engineering, University of Pennsylvania, Philadelphia, PA

Sean Engles, Chenical and Biomolecular Engineering, University of Pennsylvania, Philadelphia, PA

Randall Toy, Bioengineering, University of Pennsylvania, Philadelphia, PA

Ravi Radhakrishnan, Department of Bioengineering, University of Pennsylvania, Philadelphia, PA

We will focus on applications of multiscale modeling to clathrin-dependent endocytosis, i.e., active transport mechanisms characterized by vesicle nucleation and budding of the cell membrane orchestrated by protein-interaction networks. We will discuss theoretical and computational methodologies for quantitatively describing how cell-membrane topologies are actively mediated and manipulated by intracellular protein assemblies. We will describe the development and application of the kinetic Monte Carlo time-dependent Ginzburg Landau (KMC-TDGL) algorithm for unified dynamics of curvature-inducing proteins and membranes. We will also describe a minimal model that captures the design principles behind the self-assembly and adaptive morphogenesis of the clathrin coat. Together, these models provide mechanistic insight into the "hidden" traffic rules in intracellular trafficking mechanisms.

See more of this Session: In Silico Systems Biology: Intracellular Signaling and Gene Regulation
See more of this Group/Topical: Topical A: Systems Biology

A Dynamic Analysis of Insulin Signaling and Its Feedback Mechanisms: A Discrete Modeling Approach

Ming Wu, Computer Science, Michigan State University, East Lansing, MI

Xuerui Yang, Chemical Engineering and Materials Science, Biochemistry and Molecular Biology, Michigan State University, East Lansing, MI

Christina Chan, Chemical Engineering and Materials Science, Biochemistry and Molecular Biology, Michigan State University, East Lansing, MI

A major challenge in systems biology is to understand the dynamic behavior of biological signaling systems, which depends on the elements of the system and their inter-relationships. Computational modeling plays an integral part in the study of network dynamics and uncovering the underlying mechanisms. However, it is difficult to acquire quantitative measurements of all the biochemical and kinetic parameters of the signaling pathways. Thus, an alternative modeling method is needed that effectively integrates pathway information and explores the basic dynamics of the signaling network, in the absence of detail kinetic parameters.

Here we applied a discrete dynamic modeling approach using three-state logic variables to describe the protein activity levels, and transition functions in a signaling network. The model takes into account cell-cell variations on the protein activities and reaction rates, and thereby can simulate a large population of cells. The simulation results provided patterns of dynamic activation for each component in the network, despite the absence of detailed kinetic parameters and uncertainties within the signaling process (i.e., cell-cell variation in reaction timing or protein activity levels).

Our dynamic model of the insulin signaling system in liver cells provides a proof-of-concept application of the proposed methodology. Research in our group has shown that the double-stranded RNA-dependent protein kinase (PKR), which is affected by insulin, is involved in regulating insulin signaling through a feedback mechanism. As a downstream target of insulin receptor substrate 1 (IRS1), PKR induces the inhibitory serine phosphorylation of IRS1, thereby suppressing its tyrosine phosphorylation. In fact, the dynamic behavior of the insulin signaling network is tuned by a variety of feedback pathways, many of which have the potential to cross-talk with PKR, but the contribution of each is unclear. Given the complexity of insulin signaling, it is inefficient to experimentally test all possible interactions in the network to confirm their functionality. Our discrete dynamic model provides an *in silico* model framework that integrates potential interactions and assesses the contributions of the various interactions on the dynamic behavior of the signaling network. Simulations with the model generated testable hypothesis on the response of the network upon perturbation, which were experimentally evaluated to identify the pathways that actually are functioning in the particular liver cell system. Our modeling study in combination with the experimental results, elucidated the feedbacks involved, which suggested the insulin network is robustly designed.

Thus, our discrete modeling methodology, in spite of its simplicity, is able to reproduce the dynamic profiles, provide testable hypotheses, and help to analyze the major characteristics of the insulin signaling network. Such a modeling framework that relies solely on the network architecture can be easily extended to other dynamic network systems and serves as a basis to guide model-based experiments and potentially more detailed inquiry into the regulatory mechanisms of biological networks.

See more of this Session: In Silico Systems Biology: Intracellular Signaling and Gene Regulation
See more of this Group/Topical: Topical A: Systems Biology

Emergent Bistability by a Growth-Modulating Positive Feedback Circuit

Chee Meng Tan, Biomedical Engineering & Institute for Genome Sciences and Policy, Duke University, Durham, NC

Philippe R. Marguet, Dept. of Biochemistry, Duke University, Durham, NC

Lingchong You, Biomedical Engineering, Duke University, Durham, NC

A synthetic gene circuit is often engineered by considering the host cell as an invariable "chassis". Circuit activation, however, may modulate host physiology, which in turn can drastically impact circuit behavior. We illustrate this point by a simple circuit consisting of mutant T7 RNA polymerase (T7 RNAP*) that activates its own expression in bacterium Escherichia coli. Although activation by the T7 RNAP* is noncooperative, the circuit caused bistable gene expression. This counterintuitive observation can be explained by growth retardation caused by circuit activation, which resulted in nonlinear dilution of T7 RNAP* in individual bacteria. Predictions made by a model accounting for such effects were verified by further experimental measurements. Our results reveal a novel mechanism of generating bistability and underscore the need to account for host physiology modulation in engineering gene circuits.

See more of this Session: Synthetic Systems Biology I
See more of this Group/Topical: Topical A: Systems Biology

Bistable Gene Regulatory Networks: Intrinsic Versus Externally Induced Bistability

John S. F Barrett, Chemical Engineering and Material Science, University of Minnesota, Minneapolis, MN

Jennifer A. Maynard, Chemical Engineering, University of Texas, Austin, Austin, TX

Yiannis N. Kaznessis, Chemical Engineering and Materials Science, University of Minnesota, Minneapolis, MN

In this paper we bring light to the topic of bistability in gene regulatory networks by making the important distinction between intrinsically bistable systems which arise solely from the balance of gene expression between two cross-repressing operons and those systems which are actually monostable but become bistable in the presence of an external inducer molecule. This property of a bistable gene network can affect how it is suited for a specific function in nature and how synthetic bistable systems can be custom designed for a particular application in biotechnology. Finally we clarify the meaning of bistability in the context of stochasticity as this phenomenon blurs the distinction between monostable and intrinsically bistable phenotypes and influences the long term stability of externally induced bistable systems.

For our study we have constructed a simple gene network in *Escherichia coli* comprised of two genetic operons carried on artificial plasmids. One operon consists of a promoter that controls the expression of the tetracycline-inducible repressor protein (TetR) and the red fluorescent, reporter protein (Mcherry). The promoter is regulated by the lactose-inducible repressor protein (LacI). The other operon consists of a TetR-repressible promoter that controls expression of the LacI protein and the green fluorescent, reporter protein (GFPmut2). Acting together, the two operons produce a bimodal phenotype in which each population is dominated by a different operon. To control the distribution of phenotypes within the population we use externally added, inducer molecules, IPTG and aTc, capable of binding LacI and TetR, respectively. In the induced state the repressor molecules undergo an allosteric transition resulting in reduced affinity for their target operator regions, thus promoter repression is relieved.

A very similar LacI / TetR based device, named "the genetic toggle switch" was first constructed in 2000 by Gardner et al.[1] Beyond their work we highlight the need for a more rigorous definition of genetic bistability: 1.) when external inducer molecules are used in the system and 2.) when the system is subject to stochastic fluctuations of relatively high probability. Through *in vivo* experiments and *in silico* stochastic simulations we demonstrate that this particular network configuration is actually capable of at least 4 different classes of behavior each exhibiting a different form of bistability. These classes include: intrinsic bistability, IPTG induced bistability, aTc induced bistability, and a fourth class in which cells are induced by either IPTG or aTc. This more extensive classification will add to the range of application for bistable gene networks.

By adjusting repressor half-life through ssrA degradation tags, we constructed 4 different system variants to explore the different classes of bistable behavior. In addition to results from flow cytometry showing the distribution of phenotypes for each system variant, we also present *in silico* model data to demonstrate the benefit of highly mechanistic stochastic kinetic models for modeling gene regulatory networks and generating useful predictions for molecular and synthetic biologists.

1. Gardner TS, Cantor CR, Collins JJ. 2000. Construction of a genetic toggle switch in *Escherichia coli*. Nature 403(6767):339-42.

See more of this Session: Synthetic Systems Biology I
See more of this Group/Topical: Topical A: Systems Biology

A Synthetic Approach to Understanding Signal Processing in MAPK Cascades

Ellen C. O'Shaughnessy, Department of Biomedical Engineering, Boston University, Boston, MA

Santhosh Palani, Department of Bioengineering, University of Pennsylvania, Philadelphia, PA

James J. Collins, Department of Biomedical Engineering, Boston University, Boston, MA

Casim A. Sarkar, Department of Bioengineering, University of Pennsylvania, Philadelphia, PA

The mitogen-activated protein kinase (MAPK) cascade is a ubiquitous and versatile signaling module found in all eukaryotic cells. MAPK cascades consist of three-tiered phosphorylation events and are essential to various biological processes such as differentiation, proliferation, motility, and apoptosis. They exhibit a broad array of system dynamics and these dynamics are often crucial in determining the biological outcome of activation. However, analysis of signal processing in endogenous mammalian MAPK cascades can be complicated by signal crosstalk and by difficulties in perturbing or rewiring the network topology in a mammalian cell. We have employed a synthetic biology approach to build an insulated mammalian MAPK cascade in yeast, and we have explored how variations in topology can influence the response of this signaling module. We will describe how the dynamics of a basic synthetic MAPK cascade can be modulated by incorporating scaffold proteins, by adding negative regulation, or by varying the concentration of the cascade. Our results indicate that the MAPK cascade is an inherently flexible signal-processing module and also elucidate mechanisms of signal tuning that may be more generally useful for probing and engineering system dynamics.

See more of this Session: Synthetic Systems Biology I
See more of this Group/Topical: Topical A: Systems Biology

Tuning Gene Expression Hetereogeneity with Cis-Encoded Epigenetic Switches

Narendra Maheshri, Chemical Engineering, MIT, Cambridge, MA

Leah Octavio, Computational and Systems Biology, MIT, Cambridge, MA

Epigenetic switches encode their state information either locally, often via covalent modification of DNA or histones; or globally, usually in the level of a trans-regulatory factor. Here we examine how the regulation of local epigenetic switches controls the extent of heterogeneity in gene expression, which is ultimately tied to phenotypic diversity in a population. We show that two copies of a locally encoded epigenetic switch, the *FLO11* locus in *S. cerevisiae*, switch between a silenced and competent promoter state in a random and independent fashion. Using a stochastic kinetic model, we further quantify the effect trans regulators have on the slow and fast transitions between promoter states associated with epigenetic and conventional regulation of *FLO11*. We find different classes of regulators affect epigenetic, conventional, or both forms of regulation. Distributing the task of epigenetic and conventional regulation represents a potent strategy that offers cells flexibility in shaping the distribution of gene expression within a population. We further investigate the basic architecture of the *FLO11* promoter to understand the molecular basis of the stability of the silenced and competent promoter states. A systems-level analysis of this complex promoter reveals basic strategies for increasing stability similar to signal-transduction networks. Finally, we discuss possible applications of local, cis-encoded epigenetic switches like *FLO11* in synthetic biology and metabolic engineering.

See more of this Session: Synthetic Systems Biology I
See more of this Group/Topical: Topical A: Systems Biology

A One-Step Procedure for the Optimal Connection of Synthetic Genetic Circuits

Howard Salis, Department of Pharmaceutical Chemistry, UC San Francisco, San Francisco, CA

Christopher Voigt, Pharmaceutical Chemistry, University of California San Francisco, San Francisco, CA

Engineering complex genetic circuits remains a time-consuming and laborious affair. These systems combine multiple genetic elements, such as promoters, ribosome binding sites, and protein coding sequences, to create interesting and potentially useful dynamical or logical behaviors. While individual genetic circuits may function as expected, successfully connecting them together to create higher-order behaviors often requires significant trial-and-error mutagenesis. Connections are made by having the output of one genetic circuit, typically a transcription factor, regulate the input of another genetic circuit. A key challenge is determining the production rates of the inputs and outputs such that their overall regulatory behavior exhibits the desired dynamics or logic.

Here, we present a formal design procedure for functionally connecting synthetic genetic circuits together, requiring only a single genetic modification to the original system. First, the method identifies the optimal production rates of the inputs and outputs of two genetic circuits, such that the systems behavior exhibits a desired logical behavior. Second, the procedure generates a specific DNA sequence that will enact these changes. To do this, the design procedure has three ingredients: (i) the computational predictions of the ribosome binding site (RBS) calculator[1], a newly developed predictive design method for ribosome binding sites; (ii) the transfer function data for each synthetic genetic circuit, quantifying the steady-state relationship between their inputs and outputs; and (iii) the original DNA sequence of the two synthetic genetic circuits. The user defines a fitness function describing the desired logical behavior and the procedure uses these three ingredients to identify the optimal sequence of the regulatory protein's ribosome binding site that maximizes the value of the fitness function. Inserting this ribosome binding site sequence upstream of the protein will result in the successful connection of two synthetic genetic circuits with an optimal higher-order behavior.

To demonstrate its utility, we employ the design procedure to functionally connect together two synthetic genetic circuits. As an example, we connect a pair of signal-responsive inputs to a previously constructed AND logic circuit[2]. Without genetic modification, the production rates of the inputs are either too high or too low, causing the output of the logic circuit to poorly exhibit AND logic. Given a fitness function describing AND logic, the design procedure predicts the sequence of a ribosome binding site such that the correct AND logic is outputted. We experimentally verify that the generated sequence results in near-optimal AND logic by characterizing the combined system. We also rationally perturb the input rates to measure the systems's deviation from optimality.

[1]Salis H, Mirsky EA, Voigt CA, "Automated Design of Synthetic Ribosome Binding Sites to Precisely Control Protein Expression", (in review)

[2]Anderson JC, Voigt CA, Arkin AP, "Environmental signal integration by a modular AND gate", Nat. Mol. Syst. Biol., 2007

See more of this Session: Synthetic Systems Biology I
See more of this Group/Topical: Topical A: Systems Biology

Bottom up Engineering of Synthetic Gene Networks

Xiao Wang, Biomedical Engineering Department, Boston University, Boston, MA

Tom Ellis, Biomedical Engineering Department, Boston University, Boston, MA

James J. Collins, Biomedical Engineering Department, Boston University, Boston, MA

Synthetic gene networks can be constructed from bottom up with desired properties. However, constructing predictable gene networks with desired functions remains a challenge. It is because of the lack of well-characterized components and unpredictability of the assembled networks. The need for extensive, iterative characterization and retrofitting for optimization drastically slows down the construction of desired synthetic gene networks. Here we present a diversity-based approach that combines promoter library synthesis and mathematical modeling to quickly construct gene networks with desired properties (1). In this approach, promoters with random strength diversities are synthesized and characterized in parallel. When coupled with mathematical modeling to simulate the network at a whole system level, promoters that are optimal for the intended functions can be selected before the actual network assembly, without the need for post-hoc modifications. This approach will first be demonstrated in yeast by constructing negative feedforward loop networks. Then the method will be used to produce a synthetic gene network that acts as a timer, tunable by component choice. We utilize this network to control the timing of yeast flocculation phenotype, to illustrate a practical application of our approach. The construction of a synthetic cellular counter (2) will also be presented to lead to the discussion of future directions for library-modeling based approaches in synthetic biology.

1. Ellis, T.*, X. Wang*, and J. J. Collins. 2009. Diversity-based, model-guided construction of synthetic gene networks with predicted functions. Nature biotechnology 27:465-471.

2. Friedland, A. E.*, T. K. Lu*, X. Wang, D. Shi, G. Church, and J. J. Collins. 2009. Synthetic gene networks that count. Science 324:1199-1202.

(* equally contributed)

See more of this Session: Synthetic Systems Biology I
See more of this Group/Topical: Topical A: Systems Biology

Enhancing Production of Glucaric Acid From a Synthetic Pathway in Recombinant Escherichia Coli

Tae Seok Moon, Chemical Engineering, Massachusetts Institute of Technology, Cambridge, MA

John E. Dueber, California Institute of Quantitative Biomedical Research, University of California, Berkeley, Berkeley, CA

Sang-Hwal Yoon, Chemical Engineering, Massachusetts Institute of Technology, Cambridge, MA

Eric Shiue, Chemical Engineering, Massachusetts Institute of Technology, Cambridge, MA

Kristala Jones Prather, Chemical Engineering, Massachusetts Institute of Technology, Cambridge, MA

Key features of Synthetic Biology include a focus on design and design principles, as well as the development of well-characterized and re-usable Parts. The field intersects with Metabolic Engineering in areas such as the design of novel pathways for product generation, in which enzymes may be considered as interchangeable Parts, and the improvement of those pathways for increased productivity. We have constructed a synthetic pathway for the production of glucaric acid, deemed a "top-value added chemical" from biomass, from glucose in Escherichia coli. Co-expression of the genes encoding myo-inositol-1-phosphate synthase (Ino1) from Saccharomyces cerevisiae, myo-inositol oxygenase (MIOX) from mouse, and uronate dehydrogenase (Udh) from Pseudomonas syringae led to production of glucaric acid. Flux towards glucaric acid is ultimately limited by MIOX, whose activity is dependent upon the concentration of myo-inositol, its substrate. To improve glucaric acid production, we have explored several options for increasing flux through the pathway, including the use of other enzyme Parts and the creation of synthetic scaffolds (Devices) to co-localize Ino1 and MIOX, thereby increasing the local concentration of myo-inositol. We will present experimental results on the application of these scaffolds in various configurations to improve MIOX activity and glucaric acid productivity. We will also present numerical simulations that attempt to explore the parameters for re-usability of these Devices and explain observed differences in the behavior of the scaffolds as applied here compared to a previously explored system.

See more of this Session: Synthetic Systems Biology II
See more of this Group/Topical: Topical A: Systems Biology

Will Cell-Free Synthetic Pathway Biotransformation (SyPaB) Compete with Microbial Fermentation for Producing Biocommodities?

Y.-H. Percival Zhang, Biological Systems Engineering Department, Virginia Polytechnic Institute and State University, Blacksburg, VA

A single enzyme can catalyze a one-step chemical reaction, while a microorganism can transform a complicated chemical reaction by integrating numerous enzymatic steps. In vitro (cell free) synthetic biology projects are a largely unexplored strategy. Different from most of cell-free synthetic biology projects that produce high-value biopolymers (e.g., proteins) from small building bricks, cell-free synthetic pathway biotransformation (SyPaB) is to in vitro assemble a number of enzymes and coenzymes for producing low-value products (e.g., hydrogen) by breaking down natural biopolymers (e.g., cellulose and starch). Recently, novel synthetic pathways composed of 13-14 enzymes have been demonstrated to produce nearly 12 molecules of hydrogen per molecule of glucose unit of starch or cellulosic materials and water as $C_6H_{10}O_5$ (aq) + 7 H_2O (l) --> 12 H_2 (g) + 6 CO_2 (g) [1,2]. Here we will further analyze the technological and economical feasibility of SyPaB for producing low-value high-volume biocommodities. SyPaB is expected to have comparative advantages over microbial fermentations, such as high product yield, fast reaction rate, high product titer, simple product separation, easy control, and so on.

[1] Y.-H. P. Zhang, B. R. Evans, J. R. Mielenz, R. C. Hopkins, M. W. W. Adams, PLoS One 2007, 2, e456.

[2] X. Ye, Y. Wang, R. C. Hopkins, M. W. W. Adams, B. R. Evans, J. R. Mielenz and Y.-H. P. Zhang, ChemSusChem, 2009, 2, 149-152.

See more of this Session: Synthetic Systems Biology II
See more of this Group/Topical: Topical A: Systems Biology

Using Synthetic Metagenomics to Discover Biocatalysts for Fuels Synthesis

Travis S. Bayer, University of California, San Francisco, San Francisco, CA

The conversion of biomass to next generation fuels and chemicals will require the discovery and engineering of novel enzymes and metabolic pathways. We used a synthetic biology and metagenomics approach to discover enzymes that catalyze the formation of methyl halides, a reactive carbon building block that can be chemically converted to fuels and chemicals. A single methyl halide transferase (MHT) enzyme transfers the methyl group from the ubiquitous metabolite S-adenoyl methionine (SAM) to a halide ion. Plants and microorganisms naturally produce methyl halides, but these organisms produce very low yields and are not amenable to industrial production. We used bioinformatics and chemical synthesis to identify and construct a library of 89 putative genes from sequence data in the NCBI database. The set was screened in Escherichia coli to identify the rates methyl halide production. We transferred the highest activity gene to yeast and re-engineered SAM pathway regulation to dramatically increase methyl halide productivity and conversion efficiency. Using a symbiotic co-culture of the engineered yeast and the cellulolytic bacterium Actinotalea fermentans, we are able to achieve methyl halide production from unprocessed switchgrass, corn stover, sugarcane bagasse, and poplar. These results demonstrate the potential of producing methyl halides from non-food agricultural resources, and highlight strategies for discovering and optimizing novel routes for converting biomass to fuels and chemicals.

See more of this Session: Synthetic Systems Biology II
See more of this Group/Topical: Topical A: Systems Biology

Towards Construction of a Synthetic Self-Replicating Entity

Michael C. Jewett, Genetics, Harvard Medical School, Boston, MA

A powerful approach to enhance our understanding of complex biomolecular systems is to build integrated biological processes from their fundamental components. The long-term vision of this project is to construct a useful synthetic entity that self-replicates from small molecule building blocks and reconstitutes the macromolecular catalysts synthesizing DNA, RNA and protein. This objective demands in vitro methods for production and self-assembly of active component parts. Here we describe recent efforts towards constructing highly active E. coli ribosomes from their constituent parts. Previous works have shown total reconstitution of functionally active 30S and 50S subunits from native E. coli rRNA and proteins. However, reconstituted ribosomes have only ever been shown competent for their ability to translate poly-uridine [Poly(U)] mRNA templates into polyphenylalanine. Moreover, reconstituting E. coli ribosomes under physiological conditions has remained elusive for decades. Here we show that reconstituted ribosomes can synthesize active full-length protein, luciferase and green fluorescent protein. Furthermore, we show that active E. coli ribosomes can be reconstituted in a one-step incubation procedure at 37°C under conditions that mimic the cytoplasm. Finally, as a milestone towards building a synthetic replicon, we provide evidence that rRNA synthesis, ribosome assembly, and protein synthesis can be carried out in one compartment. Our reconstitution methods will contribute towards the in vitro analysis of ribosomal mutations for understanding the molecular details of ribosome function. Equally important, our approach solves one of the critical barriers to constructing ribosomes in vitro and promises to enable cell-free synthetic biology as a platform for evolving ribosomes for the production of protein therapeutics and peptide drugs that are difficult to make in vivo because of their toxicity, complexity, or unusual cofactor requirements.

See more of this Session: Synthetic Systems Biology II
See more of this Group/Topical: Topical A: Systems Biology

Design and Construction of a Synthetic, Tunable Microbial Consortium

Alissa R. Kerner, Chemical Engineering, University of Michigan, Ann Arbor, MI

Xiaoxia Lin, Chemical Engineering, University of Michigan, Ann Arbor, MI

Microbial symbiosis confers many advantages to its participants and has the potential to extend beyond naturally occurring systems. Applications for this kind of synthetic system include microbial production of molecules that are difficult to synthesize or may take many different steps and/or reactions to produce, such as biofuels or pharmaceuticals. Here we attempt to create a tunable microbial symbiotic system by genetically engineering E. coli auxotrophs to cross-feed and support each other when grown in co-culture. Synthetic metabolic circuits were designed to allow tunable control of the system via the export of essential amino acids by each of the auxotrophs. In our system, a tyrosine auxotroph (Δtyr) produces tryptophan for the tryptophan auxotroph (Δtrp) and vice versa. Inducible promoters are used to control the expression of relevant target genes, which then control the amino acid synthesis and export. It has been previously shown that the growth rate and composition of such a co-culture are determined by the export rate and cellular requirement of essential molecules. By adding specific amounts of inducers, we have been able to produce a gradient of target gene expression and a co-culture with variable growth rate and composition.

See more of this Session: Synthetic Systems Biology II
See more of this Group/Topical: Topical A: Systems Biology

Engineering Human T-Cell Receptor Signaling Dynamics

Wilson W. Wong, Cellular and Molecular Pharmacology, University of California, San Francisco, San Francisco, CA

Ethan Corcoran, Microbiology and Immunology, University of California, San Francisco, San Francisco, CA

Arthur Weiss, Microbiology and Immunology, University of California, San Francisco, San Francisco, CA

Wendell A. Lim, Cellular and Molecular Pharmacology, University of California, San Francisco, San Francisco, CA

T-cells are important regulators of the immune response. Central to the immune response of T-cells is their activation through the T-cell receptor (TCR). Although under intense investigation, how the duration and level of TCR activation affects T-cell response, such as IL-2 production and T-cell proliferation, remains unclear. Recent studies have demonstrated the power of synthetic biology approaches for understanding the design principles of signaling systems; by engineering novel circuit architectures one can probe how diverse dynamic behaviors can emerge from a common set of core signaling components. Here we used the synthetic approach to create new feedback loops and TCR signaling dynamics, and ask how the different dynamics control the ultimate physiological response. By creating libraries of circuits, we can systematically map what architectures and parameters yield particular dynamic behaviors, and the physiological consequences. The discrepancy between expectations and reality will also provide us focal points to explore our knowledge gap. Moreover, this work can complement ongoing works on therapeutic T-cells engineering by creating user-defined temporal and dose-response behaviors. Furthermore, this work will yield a new set of genetic tools to control the TCR signaling dynamics that can benefit the signal transduction community.

See more of this Session: Synthetic Systems Biology II
See more of this Group/Topical: Topical A: Systems Biology

A Multi-Scale Model for the Assessment of Autonomic Dysfunction in Human Endotoxemia

P.T. Foteinou, Biomedical Engineering, Rutgers University, Piscataway, NJ

S.E. Calvano, Department of Surgery, UMDNJ-Robert Wood Johnson Medical School, New Brunswick, NJ

S.F. Lowry, Surgery, UMDNJ-Robert Wood Johnson Medical School, New Brunswick, NJ

Ioannis P. Androulakis, Biomedical Engineering, Chemical & Biochemical Engineering, Rutgers University, Piscataway, NJ

Severe injury and infection are leading causes of morbidity and mortality evoking both pro-inflammatory and anti-inflammatory responses at the systemic and tissue levels[1]. Mechanisms aiming at the regulation of the inflammatory response involve not only the local release of anti-inflammatory cytokines but also hormonal and autonomic influences. Recent studies indicate that the central nervous system (CNS) is a pivotal regulator of the immune response and its autonomic divisions control inflammation at various levels. In particular, the neuro-immune bidirectional network is comprised of a descending pathway that links CNS to peripheral immune tissues and a parallel afferent arm linking the immune system with the CNS[2]. The integrity of this loop allows for communication between the CNS and the peripheral immune system, integrating neuronal and immune signals in the periphery as well as in the CNS. Such reciprocal communication pathways between the immune and central nervous systems are now considered to be major components of the integrated homeostatic network of the host and involve all areas of medulla, hypothalamic regions and autonomic system[3]. It is therefore becoming evident that disruptions may occur at any level of these adaptive functions and thereby predispose a host to excessive inflammatory responses.

Heart rate variability (HRV), assessed by evaluating the standard deviation of normal to normal interbeat intervals, is a noninvasive means of quantifying the cardiac autonomic input. Assessment of HRV has been extensively used to evaluate autonomic modulation of the sinus node and as a predictor of the severity of illness[4]. It has been hypothesized that the reduction in HRV, i.e., increases in regularity, suggests an increased isolation of the heart with other organs that behave like biological oscillators coupled to one another. Loss of high level signal variability reflects detectable systemic-level loss of adaptability and fitness[5]. As a potential surrogate marker for system decomplexification, diminished HRV and autonomic dysfunction have received increasing attention in the mathematical modeling of critical illness. Thus, the development of surrogate multi-scale, nonlinear, dynamic, complex models gains ever increasing acceptance as a means of deciphering the intricacies of critical illness emphasizing the emerging loss ("decomplexification") of rich variability, observed in normal physiology, in critically ill patients[6].

Understanding the relevance of neuro-immunomodulation in controlling inflammatory processes would require a model that simulates critical aspects of the multi-scale interplay between the central nervous system (CNS) and the immune response. A vital enabler in that respect is the development of a systems-based approach that integrates clinical data across multiple scales and subsequently models the emerging host response as the outcome of orchestrated interactions of critical modules. To meet this challenge, an integrated host response model is proposed to connect the (cellular) inflammatory response with neural based pathways that account for modulations in autonomic activity assessed by physiologic variables including HRV. Specifically, the data used in this study span various scales from the cellular to the systemic host response level.

Based on our prior work[7-9], a physicochemical model of human endotoxemia, as a prototype model of systemic inflammation in humans, will be used as template for connecting extracellular signals and intracellular signaling cascades eventually leading to the emergent transcriptional dynamics. However, one of the simplifying hypotheses made in this model is that the immunomodulatory role of autonomic (hormonal) influences is not accounted for. Driven by the premise that a characteristically enhanced endocrine hormone profile is elicited during the early-phase response to endotoxin injury[10], the additional scale we wish to explore is associated with critical aspects of the neuro-endocrine immune crosstalk. In particular, essential modules associated with the release of neuro-endocrine stress hormones including endogenous cortisol and catecholamines are further incorporated. Accordingly, a signal transduction model is developed to establish implicit interaction among these signaling molecules and thereby quantify the bi-directional link between the neuro-endocrine axis and peripheral inflammation. Clinical observables such as HRV will be further incorporated to assess disruptions in autonomic activities which reflect the manifestation of systemic decomplexification and correlate with the severity of the host. In order to quantify the relevant system parameters and functions, data associated with plasma cortisol and catecholamine levels recorded before and after the endotoxin infusion are employed. Of particular relevance are also data associated with vital signs including heart rate and mean arterial blood pressure. Such modeling effort will be evaluated through its ability to not only reproduce available experimental data but also to assess the implications of endocrine hormone influence upon autonomic activity. Thus, the proposed work intends to associate acquired endocrine dysfunction with diminished HRV making it a critical enabler for the assessment of neuro-endocrine activity at both the cellular and systemic (autonomic) levels[11].

The overall goal of this study is to demonstrate the feasibility of a clinically relevant, mechanistic-based, multi-level human inflammation model that couples essential aspects of the bi-directional relationship between the CNS and inflammation. Specifically, the proposed modeling effort bridges the initiating signal and macroscopic phenotypic expressions (HRV) through semi-mechanistic models that include transcriptional dynamics, signaling cascades and physiological components. Such systems level approach could potentially offer significant insight on how interacting inflammatory responses and neural based mechanisms influence the host's ability to regulate inflammation; thus making it a critical enabler for clarifying the clinical contexts in which disruptions of these bidirectional communications contribute to morbidity and mortality to severely stressed patients.

1. Lowry, S.F., The stressed host response to infection: the disruptive signals and rhythms of systemic inflammation. Surg Clin North Am, 2009. 89(2): p. 311-26.

2. Pavlov, V.A. and K.J. Tracey, Neural regulators of innate immune responses and inflammation. Cell Mol Life Sci, 2004. 61(18): p. 2322-31.

3. Taub, D.D., Neuroendocrine interactions in the immune system. Cell Immunol, 2008. 252(1-2): p. 1-6.

4. Norris, P.R., et al., Heart rate variability predicts trauma patient outcome as early as 12 h: implications for military and civilian triage. J Surg Res, 2005. 129(1): p. 122-8.

5. Seely, A.J. and N.V. Christou, Multiple organ dysfunction syndrome: exploring the paradigm of complex nonlinear systems. Crit Care Med, 2000. 28(7): p. 2193-200.

6. Buchman, T.G., Nonlinear dynamics, complex systems, and the pathobiology of critical illness. Curr Opin Crit Care, 2004. 10(5): p. 378-82.

7. Foteinou, P.T., et al., An indirect response model of endotoxin-induced systemic inflammation. Journal of Critical Care, 2007. 22(4): p. 337-338.

8. Foteinou, P.T., et al., Modeling endotoxin-induced systemic inflammation using an indirect response approach. Math Biosci, 2008. 217(1): p. 27-42.

9. Foteinou, P.T., et al., In silico simulation of corticosteroids effect on an NFkB- dependent physicochemical model of systemic inflammation PLoS ONE, doi: 10.1371/journal.pone.0004706 2009 4(3).

10. Lowry, S.F., Human endotoxemia: a model for mechanistic insight and therapeutic targeting. Shock, 2005. 24 Suppl 1: p. 94-100.

11. Alvarez, S.M., et al., Low-dose steroid alters in vivo endotoxin-induced systemic inflammation but does

not influence autonomic dysfunction. J Endotoxin Res, 2007. 13(6): p. 358-68.

See more of this Session: Multiscale Systems Biology
See more of this Group/Topical: Topical A: Systems Biology

How Antigen Quality and Quantity Determine T Cell Triggering In Vivo

Huan Zheng, Chemical Engineering, MIT, Cambridge, MA

Arup Chakraborty, Biological Engineering, Chemical Engineering and Chemistry, MIT, Cambridge, MA

Search and recognition of invading antigen presented by the peptide major histo-compatibility complex (pMHC) on the surface of dendritic cells in lymph nodes precede the triggering of T cell receptor (TCR) signaling and the subsequent T cell activation. Using computer simulations aided by experimental studies, we studied signaling events as T cells migrate through the lymph node searching for cognate antigen. We showed that the antigen dosage and quality determine whether and when T cells stop migrating and engage dendritic cells in prolonged conjugation to achieve full activation. Our theoretical analyses of the in silico results further identified a consolidated measure of antigen quantity and type, using which the effects of various measures of antigen quantity and quality on the degree of T cell effector response can be collapsed onto one single curve.

Very few cognate pMHC are needed to trigger TCR signaling and effector functions in Vivo. This suggests that T cells may integrate sub-optimal signals from sequential transient contacts with dendritic cells. Such signal integration, however, would require a mechanism for short-term molecular "memory". Recent computational and experimental studies have revealed that positive feedback regulation of membrane-proximal signaling (Ras activation) causes previously stimulated cells to be rapidly triggered upon subsequent weak re-stimulation. We developed a computational model that couples T cell migration with membrane proximal signaling activities triggered by encounters with dendritic cells bearing pMHCs of different affinities, in order to examine the conditions for signal integration to occur via this mechanism. Specific predictions are made regarding how to test whether this mechanism enables signal integration, and its consequences for T cell sensitivity and the role of endogenous peptides during the earliest events in detecting infectious pathogens.

See more of this Session: Multiscale Systems Biology
See more of this Group/Topical: Topical A: Systems Biology

Model-Based Analysis and Quantitative Measurement of Key Components of Tumor Necrosis Factor Trafficking in a Tuberculosis Granuloma

Mohammad Fallahi-Sichani, Chemical Engineering, University of Michigan, Ann Arbor, MI

Matthew A. Schaller, Pathology, University of Michigan Medical School, Ann Arbor, MI

Denise E. Kirschner, Microbiology and Immunology, University of Michigan Medical School, Ann Arbor, MI

Steven L. Kunkel, Pathology, University of Michigan Medical School, Ann Arbor, MI

Jennifer J Linderman, Chemical Engineering, University of Michigan, Ann Arbor, MI

Tuberculosis (TB) is caused by a highly successful bacterium, *Mycobacterium tuberculosis* (Mtb), and is responsible for 2 million deaths per year. The interplay between host and bacterial factors at a variety of length and time scales leads to different disease outcomes (primary disease, latency, or TB reactivation). A key outcome of TB infection is the formation of a granuloma. Tuberculosis granulomas are organized collections of immune cells composed of macrophages, lymphocytes and other cells that form in the lung in response to Mtb infection. Granuloma formation is essential for immunologic and physical control of mycobacterial infection. Maintenance of granulomas is required during latent TB, during which infection is contained with no symptoms. Our overall hypothesis is that events at several spatial (molecular, cellular, tissue, body) and time (minutes to years) scales affect the development and maintenance of TB granulomas, and that a model of granuloma formation that encompasses multiple scales is necessary to fully understand the immune response to Mtb. In this work, we focus specifically on molecular, cellular, and multi-cellular aspects of the granuloma and the central role of tumor necrosis factor α (TNF) in these interactions.

TNF is well-known as a key effector molecule in the host protective immune response against tuberculosis. This cytokine can be produced by immune cells including macrophages, neutrophils, and T cells and exerts its action on multiple cell types by affecting cell activation and migration, expression of chemokines, apoptosis and other biological processes. Numerous reports have revealed the role of TNF in granuloma formation, granuloma maintenance and the containment of Mtb infection [1-3]. Granuloma formation in mice that lack TNF or the TNF receptor 1 has been reported to be aberrant or delayed. Neutralization of TNF in mice with chronic infection leads to disorganization of the granulomas and subsequent death.

Our hypothesis is that TNF availability within granulomas is crucial in granuloma formation, functionality and integrity and thus control of TB infection. TNF availability is controlled at the molecular level by the kinetics of a combination of events, including TNF synthesis and release, receptor binding, internalization and degradation. Hence, to characterize the mechanisms by which TNF availability within a TB granuloma is controlled, we developed a reaction/diffusion-based differential equation model that describes a TB granuloma as a continuous collection of immune cells forming concentric layers and includes TNF-associated molecular-level events. These events include TNF synthesis by TNF-producing cells, TNF release to the extracellular space due to the activity of TNF-α converting enzyme (TACE), reversible binding of TNF to two types of TNF receptors on the cell membranes, TNF/TNF receptor intracellular trafficking, as well as TNF diffusion within granuloma. The model outcome is the fraction of TNF-bound cell surface TNF receptors in the granuloma, the major factor that controls TNF-mediated signaling and thus cellular level responses in TB infection. Parameters that significantly influence the outcome of the model were identified by sensitivity analysis. Cellular composition of granuloma, the number of TNF receptors on each cell type within granuloma, affinity of TNF receptors for binding to TNF, and the rate of TNF receptor-mediated endocytosis of TNF were shown to be the most important determinants of TNF availability within granuloma.

Experiments were performed to measure the required parameters in pulmonary granulomas induced in pre-sensitized mice by bead-immobilized mycobacterial purified protein derivative (PPD) by using multi-color flow cytometry. This experimental model has been previously demonstrated to induce an immune response typified by a type 1 cytokine phenotype which has been characterized in TB [4, 5]. PPD bead granulomas form at day 2 and reach their maximal size at day 4 following intravenous injection of PPD-coated beads. To identify the cellular composition of PPD bead granulomas and cell fractions, fluorescing antibodies for specific cell surface markers of different types of immune cells were used. Experimental results indicate that dendritic cells, macrophages, B cells, CD4+ and CD8+ T cells compose more than eighty percent of the structure of PPD-bead granulomas with B cells and macrophages showing the greatest cell fractions. A significant increase in the fraction of T cells (which represent the adaptive immune response) was observed with granuloma development up to day 4. In order to determine the rate of TNF synthesis, an inhibitor for protease activity of TACE such as TNF-α proteinase inhibitor-1 (TAPI-1) was used to inhibit TNF release from the cell membrane and the rate of TNF accumulation on the membrane was identified by quantitative flow cytometry. Dendritic cells and macrophages (at the core of granuloma) were shown to be the major TNF-producing cells within granuloma. Quantitative flow cytometry was also used to determine the cell-specific numbers of receptors within granuloma. Almost all cell types express TNF receptors, while lymphocytes down-regulate TNF receptor expression with granuloma development.

Using these experimental data on TNF-associated molecular processes and the cellular composition of PPD-bead granulomas as inputs to our multi-cellular mathematical model, we studied the role of these processes in controlling TNF availability in a TB granuloma. Our modeling results show that immune cell organization (with different levels of TNF and TNF receptor expression) within a TB granuloma is a significant factor that controls TNF availability and thus TNF signaling. Because cellular organization undergoes dynamic changes with granuloma development and at different stages of immune response (innate versus adaptive) to TB infection, it can be a factor controlling the diverse activities of TNF according to the stage of infection in the lung tissue. Cellular organization can also significantly influence the outcome of anti-TNF treatments during different stages of infection. Our modeling and experimental findings on TNF-associated molecular and multi-cellular scale aspects of granuloma can be incorporated into larger scale models describing the immune response to TB infection, including, for example, cellular input from the lymph node and circulation of immune cells in the blood. Ultimately, these modeling and experimental results can help identify new strategies for TB disease control/therapy.

[1] Flynn JL, Goldstein MM, Chan J, Triebold KJ, Pfeffer K, et al. (1995) Tumor necrosis factor-alpha is required in the protective immune response against mycobacterium tuberculosis in mice. Immunity 2: 561-572.

[2] Egen JG, Rothfuchs AG, Feng CG, Winter N, Sher A, et al. (2008) Macrophage and T cell dynamics during the development and disintegration of mycobacterial granulomas. Immunity 28: 271-284.

[3] Chakravarty SD, Zhu G, Tsai MC, Mohan VP, Marino S, et al. (2008) Tumor necrosis factor blockade in chronic murine tuberculosis enhances granulomatous inflammation and disorganizes granulomas in the lungs. Infect Immun 76: 916-926.

[4] Chensue SW, Warmington K, Ruth J, Lincoln P, Kuo MC, et al. (1994) Cytokine responses during mycobacterial and schistosomal antigen-induced pulmonary granuloma formation. production of Th1 and Th2 cytokines and relative contribution of tumor necrosis factor. Am J Pathol 145: 1105-1113.

[5] Qiu B, Frait KA, Reich F, Komuniecki E, Chensue SW (2001) Chemokine expression dynamics in mycobacterial (type-1) and schistosomal (type-2) antigen-elicited pulmonary granuloma formation. Am J Pathol 158: 1503-1515.

See more of this Session: Multiscale Systems Biology
See more of this Group/Topical: Topical A: Systems Biology

An Integrate-and-Fire Network Model to Investigate Circadian Rhythmicity in the Suprachiasmatic Nucleus

Christina Vasalou, Chemical Engineering, University of Massachusetts, Amherst, MA

Michael A. Henson, Department of Chemical Engineering, University of Massachusetts, Amherst, MA

The suprachiasmatic nucleus (SCN) of the hypothalamus is a multicellular system that drives daily rhythms in mammalian behavior and physiology. Although the gene regulatory network that drives daily oscillations within individual neurons is well characterized, less is known about the electrophysiology of the SCN cells and how firing rate correlates with circadian gene expression. We recently developed an integrate-and-fire neuron model to incorporate known electrophysiological properties of SCN pacemaker cells, including circadian dependent changes in membrane voltage and ion conductances. Calcium dynamics were included as the putative link between electrical firing and gene expression. By assuming autocrine response of the VIP and GABA neurotransmitters, the single cell model produced ion currents with oscillatory patterns matching experimental data both in current levels and phase relationships.

In this contribution, we develop SCN networks with the integrate-and-fire neuron model to analyze the relationship between network topology and synchronization behavior of electrical firing and gene expression. We build upon our previous work in which a connectivity scheme based on small world networks with both short and long range synaptic connections was used to explore VIP mediated coupling. The present model includes key neurotransmitters of the SCN (VIP, GABA and glutamate) that are believed mediate cell-to-cell communication and promote synchrony across the population. We show that VIP and GABA neurotransmitters play critical roles in daily oscillations of membrane excitability and gene expression, thereby affecting synchronicity across the network. Blocking various mechanisms of intracellular calcium accumulation by simulated pharmacological agents (nimodipine, IP3- and ryanodine-blockers) reproduced experimentally observed trends in firing rate dynamics and core clock gene transcription. We found that intracellular calcium regulates diverse circadian processes such as firing frequency, gene expression and system periodicity. The model predicts increased synchronization in the presence of sufficient cytosolic calcium levels and provides a novel multiscale framework which captures characteristics of the SCN at both the electrophysiological and gene regulatory levels.

See more of this Session: Multiscale Systems Biology
See more of this Group/Topical: Topical A: Systems Biology

Mechanochemical Energy Transduction in Muscle Contraction

Richard L. Long, New Mexico State University, Las Cruces, NM

Muscle action at the macroscale is a classic example of the integration of atomic and nanoscale events to produce macroscale action. Some details linking these scales remain in dispute. The paper by Long and Phoonsiri (CEC, 2006) shows a convincing argument for a four state acto-myosin complex model that is based on ADP release from both a high strain and low strain state, though at greatly different rates. Incorporation of strain dependence in the AMADP.Pi allows one to account for stretch as well.

See more of this Session: Multiscale Systems Biology
See more of this Group/Topical: Topical A: Systems Biology

Reprogramming Innate Immune Cells to Probe and Modulate Immunosuppressive Tumor Microenvironmental Networks

Joshua N. Leonard, Chemical and Biological Engineering, Northwestern University, Evanston, IL

Harnessing the immune system to control cancer is an attractive prospect, yet it has proven difficult to implement. Although recent technological progress has enabled such advances as patient- and tumor-specific cancer vaccinations and related immunotherapies, which have shown great promise in preclinical experiments, clinical benefits typically remain quite limited. It is now recognized that a major obstacle to the therapeutic efficacy of many such approaches is tumor-mediated immunosuppression in local microenvironments. Moreover, the mechanisms by which these multicellular networks form and are sustained remain poorly understood. Our long-term goal is to build cellular therapies that safely and effectively break down these immunosuppressive local environments in order to enhance the clinical efficacy of immunotherapies targeting a broad spectrum of cancers. To achieve this, we employ the tools of synthetic biology to build new genetic programs into cells of the innate immune system, including dendritic cells (DCs) and macrophages (MPs).

Here, we will discuss our work aimed at understanding the roles of cytokine receptor signaling and cross-talk between signaling pathways (including those of pattern-recognition receptors, such as the Toll-like receptors), in the functional polarization of innate immune cells. Both DCs and MPs are capable of taking on multiple functional roles in guiding immune function, including both immunosuppressive and immunostimulatory phenotypes. This seeming contradiction arises because these cells play important roles in both protective immunity and maintaining normal homeostasis. We employ quantitative experimental and computational tools to dissect the process by which DCs and MPs integrate a multitude of cues from their environments in order to commit to one functional polarization or another. In particular, we utilize engineered cytokine receptors and pathways, real-time and single-cell measures of gene expression, and mechanistic molecular modeling and simulation. Questions under investigation include (1) How is functional polarization determined under conditions of incoherent inputs (for example, in a mixture of pro- and anti-inflammatory cytokines)? (2) What is the role of stochastic variability in these processes? and (3) What are the requirements for sustaining or destabilizing immunosuppressive networks? This work both enhances our understanding of fundamental tumor immunology and informs the design of therapeutic interventions, which we are working to implement and evaluate.

See more of this Session: Multiscale Systems Biology
See more of this Group/Topical: Topical A: Systems Biology

MicroRNA Profiling and Discovery for Process Enhancement in Chinese Hamster Ovary Cells

Kathryn C. Johnson[1], Nitya M. Jacob[1], Matthias Hackl[2], Peter Morin Nissom[3], and Wei-Shou Hu[1]

[1]Department of Chemical Engineering and Materials Science, University of Minnesota, 421 Washington Ave. SE, Minneapolis, MN 55455
[2]Department of Biotechnology, University of Natural Resources and Applied Life Sciences, Muthgasse 18 (House B), A-1190 Vienna, Austria
[3]Bioprocessing Technology Institute, 20 Biopolis Way #06-01, Centros, Singapore 138668

Mammalian cell culture has become the predominant system used for the production of recombinant protein therapeutics. Chinese hamster ovary (CHO)-derived cell lines are the most common cell type used in industry for this purpose. The productivity of these cell lines today rivals that of professional secretor cells *in vivo*. Cell engineering via over-expression or knock-down of proteins has been an important strategy to further augment the productivity of these cell lines. Transcriptomic and proteomic studies suggest, however, that the complex traits of high productivity may result from the accumulation of changes in gene expression in many pathways. MicroRNAs (miRNAs), an abundant class of short regulatory RNAs, represent an additional level of gene expression control that could be exploited to manipulate global gene expression profiles. MiRNAs carry out simultaneous post-transcriptional regulation of up to hundreds of transcripts, and they have been implicated in the modulation of nearly all cellular processes, including cell cycle control and regulation of apoptosis.

Little work has been published on the global characterization of miRNAs in CHO cells. We show, through ultra high-throughput Illumina Solexa sequencing of seven small RNA libraries prepared from parental and recombinant CHO cell lines of varying productivities, that miRNAs are abundant in CHO cells and are highly conserved with other species. MiRNAs constituted over half of the small RNA sequencing reads, with read counts (a direct measure of transcript abundance) spanning six orders of magnitude. Hundreds of unique miRNAs were identified through sequence orthology across over two dozen different species. Many miRNAs were identified by several reads of slightly different length and sequence. These sequences, termed as isomiRs, are the result of imprecise enzymatic processing or single nucleotide extension during miRNA biogenesis. Importantly, the most abundant of these alternate sequences, often an indication of the most active forms, frequently did not correspond to the most commonly observed sequence in other species.

The sequence data also revealed differential expression of many miRNAs across the seven libraries. The time-course expression dynamics of 6 of these miRNAs were quantitatively measured in two CHO cell lines cultured under the same productivity-enhancing culture conditions represented by one of the sequenced libraries. These data were used to validate the relative miRNA expression levels obtained from the small RNA sequencing. The use of two CHO cell lines also permitted an assessment of the reproducibility of the effect of the culture conditions on miRNA expression across CHO cell lines. Potential target mRNAs were identified through the use of publically available target prediction algorithms, and pathway analysis was used to gain biological insight by identifying the major pathways being regulated. These results represent an initial assessment of the extent of miRNA influence in CHO cells, and further studies may facilitate miRNA engineering, representing a new strategy for cell engineering in mammalian cell culture.

Gene-Pair Relative Expression Binary Barcode for the Diagnosis of Various Diseases of the Human Brain

Jaeyun Sung, Chemical and Biomolecular Engineering, University of Illinois at Urbana-Champaign, Urbana, IL

Pan-Jun Kim, Institute for Genomic Biology, University of Illinois Urbana Champaign, Urbana, IL

Nathan D. Price, Chemical and Biomolecular Engineering, University of Illinois, Urbana-Champaign, Urbana, IL

One of the main challenges in cancer therapy is the accurate diagnosis of tumor class and progression grade, which can be difficult because of morphologically indistinguishable tissue representation and inter-observer variability. In order to address this issue, cancer diagnosis is moving from classification based on microscopic observations of histological specimens alone, to a method based on molecular criteria by using gene expression profiling. In this study, we have developed a disease classification scheme based on the concept of finding gene-pairs that exhibit relative expression reversals between different classes. For an extensive collection of publicly available gene expression microarray data, we have identified gene-pair relative expression decision rules that can distinguish eleven different classes and progression grades of cancerous, as well as non cancerous human brain diseases. These gene-pair decision rules were accumulated into a binary barcode panel that can serve as classifier markers for the diagnosis of brain disease. Our classifier marker panel distinguishes each disease at high accuracy according to ten-fold cross validation, proving to be a potential strategy for stratifying patients based on gene expression signatures. Furthermore, the investigation of the biological function and chromosome location of genes in the panel, led to the discovery of novel molecular properties of each brain disease, which may help elucidate some of the causes of disease manifestation. This binary barcode panel of gene-pair relative expression classifiers is the first to be built for the diagnosis of such a large number of different brain disease classes, and can serve as a highly accurate diagnostic tool in combination with traditional histopathological techniques.

See more of this Session: Genomic Approaches to Systems Biology
See more of this Group/Topical: Topical A: Systems Biology

Novel Approaches to High-Throughput Delivery of Silencing RNA

Susan T. Sharfstein, Department of Chemical and Biological Engineering, Rensselaer Polytechnic Institute, Troy, NY

Haiyuan Zhang, Department of Chemical and Biological Engineering, Rensselaer Polytechnic Institute, Troy, NY

Michael Hogg, Solidus Biosciences, Inc., Troy, NY

Moo-Yeal Lee, Solidus Biosciences, Inc., Troy, NY

Jonathan S. Dordick, Department of Chemical and Biological Engineering, Rensselaer Polytechnic Institute, Troy, NY

The development of genomic and proteomic technologies has led to the generation of enormous data sets of differentially expressed genes or proteins under varying conditions (e.g. environmental perturbations, cell differentiation, carcinogenesis, etc.) The critical issue is to identify which of these differentially expressed genes play a causal role in the various phenotypes observed under these conditions. One approach to identifying causality is the application of silencing RNA to reduce the activity of genes that are induced by the experimental system. However, silencing RNA does not readily lend itself to large-scale high-throughput analyses that would be compatible with analyzing hundreds of differentially expressed genes or proteins.

Using novel magnetic gene delivery reagents and viral transfection in combination with a proprietary three-dimensional (3D) cellular array chip (DataChip™), we have been able to transfect multiple genes and silencing RNA molecules into mammalian cells cultured in 3D arrays in 60 nL spots. Cell lines of clinical (MCF-7 breast cancer) and biotechnological (CHO-K1) relevance were transfected using this methodology. As proof of principle, cells were transfected with green fluorescent protein and the GFP was silenced using a shRNA against GFP. A variety of "toxic" shRNA molecules were also transfected and toxicity was verified by live-dead staining and analysis using a fluorescent microarray scanner. We have applied this system to investigate the responses of mammalian cells to hyperosmotic stress and evaluated the roles of five genes that were identified as differentially expressed in a microarray screen.

This system can be readily used to probe the roles of various genes in stem cell differentiation, responses to environmental stresses, or changes that occur during carcinogenesis, metastasis, etc. In addition, this system can be applied to knock down a wide range of pathways during in vitro analysis of potential pharmacological agents, permitting a mechanistic understanding of drug effectiveness.

See more of this Session: Genomic Approaches to Systems Biology
See more of this Group/Topical: Topical A: Systems Biology

Species Comparison Analysis Based On Metabolic Pathways

Meric A. Ovacik, Chemical and Biochemical Engineering, Rutgers University, Piscataway, NJ

I.P. Androulakis, Biomedical Engineering, Rutgers University, the State University of New Jersey, Piscataway, NJ

Certain fundamental remain invariable from bacteria to eukarya, therefore cross-species comparison often provides insights into underlying laws behind complex biological phenomena [1]. The general outline for the comparative analysis among species includes any similarity measure, distance matrix evaluation based on the similarity measure and phylogenic tree construction. The studies based on this approach include multiple sequence alignment of certain characteristic sequences: sequence of a single protein, single rRNA or single gene from each organism. The general standard is comparing the obtained phylogenic tree is with NCBI taxonomy [2]. The main concern about individual characteristic comparison is expressed by Felsenstein [3] that many of the conclusions derived from these studies are based on sub-parts of the phylogenetic tree and their dependability should be judged in this respect. Similarity analysis between i.e. single gene sequences cannot represent the complex relationships between species. Therefore, the whole genome is used in similarity analysis between different species. Yet, comparing genome sequences may not represent the higher level organization of organisms such as metabolic pathways and protein interaction networks.

Comparative analysis of metabolic pathways allows examining the totality of biological processes rather than the individual elements. Attempts to compare pathways include reaction content [4] and enzyme presence [5] as well as a combination of multiple features such as enzyme presence and sequence information of the enzymes in a given pathway [6-7]. The comparative analyses of pathways, however, only include individual pathways such as electron transfer, amino acid biosynthesis and TCA cycle over limited species or simpler organisms.

In this study, we expand the pathway similarity algorithm proposed by Forst et al. [7], and integrate the gene promoter similarity. We evaluate the similarity analysis for both individual pathways and over all pathways along different animal models such as rat, mouse, zebrafish, worm as well as simpler model organisms such as arabidopsis and yeast. We use the KEGG database for the pathway and the enzyme sequence information. The promoter regions are retrieved from TRANSFAC®. Over all pathways along different species, the phylogenetic relations based on our similarity measure are comparable with NCBI taxonomy; however individual pathways may differ in their own evolution context that of the organism. Our results demonstrate the potential for such an extensive bioinformatics analysis to identify appropriate surrogate species in experimental design.

[1]Wachtershauser G.(1990)Proc Natl Acad Sci U S A, 87: 200-4.

[2]Sayers EW et al. (2009) Nucleic Acids Res. D5-15.

[3]Felsenstein, J.(1988) Annu Rev Genet, 22: 521-65.

[4]Hong, S.H. et al.(2004) Appl Microbiol Biotechnol, 65(2): 203-10

[5]Heymans, M. et al. (2003) Bioinformatics,19 Suppl 1: p. i138-46.

[6]Forst, C.V. et al. (1999) Comput Biol, 6 :343-60.

[7]Forst, C.V. et al.(2001) J Mol Evol,52(6): p. 471-89.

See more of this Session: Genomic Approaches to Systems Biology

See more of this Group/Topical: Topical A: Systems Biology

Searching for Self-Renewal and Pluripotency Defining Genes in Transcriptome Space of Embryonic Stem Cells

Shikha Sharma, Department of Chemical Engineering and Materials Science, UNIVERSITY OF MINNESOTA, Minneapolis, MN

Raamesh Deshpande, Department of Computer Science and Engineering, UNIVERSITY OF MINNESOTA, Minneapolis, MN

Meri T. Firpo, Department of Medicine, Stem Cell Institute, Minneapolis, MN

Catherine Verfaillie, Stem Cell Institute and Department of Medicine, Katholieke Universiteit Leuven, Leuven, Belgium

Chad L. Myers, Department of Computer Science and Engineering, UNIVERSITY OF MINNESOTA, Minneapolis, MN

Wei-Shou Hu, Department of Chemical Engineering and Materials Science, UNIVERSITY OF MINNESOTA, Minneapolis, MN

Embryonic stem cells are characterized by their ability to self-renew indefinitely in culture and differentiate into all the three germ layers (pluripotency). The cellular state that endows these cells with these characteristics is still not well understood. Recent revelation of cellular reprogramming during the formation of induced pluripotent stem cells adds another dimension to this problem but also provides new insights into the molecular circuitry responsible for these characteristics. Microarray data from different embryonic and induced pluripotent stem cell studies was compiled from public repositories. Linear and Robust Multi-array Average normalizations were performed on the compiled datasets. Non-negative matrix factorization, a tool for factorizing high-dimensional data, was employed to identify predominant biological trends in the data. Consequently, samples were segregated into classes that were indicative of different degrees of potency. Subsequently, differential expression analysis was conducted between the predicted classes of interest using Significance Analysis of Microarrays (SAM). Functional enrichment amongst the differentially expressed genes was determined by using Metacore and Ingenuity Pathway Analysis. Genes upregulated in the pluripotent stem cells of both species (human and mouse) showed enrichment in cellular functions including DNA damage repair, G1 to S phase progression and development associated TGFβ/Activin signaling.

ARACNE, developed by Basso et al. (Nature Genetics.2005), uses gene expression profiles to identify cellular regulatory networks. This algorithm was used to predict genes whose expression is closely correlated to that of stem cell-associated genes such as Oct4, Nanog, and Sox2 etc. Genes exhibiting very similar expression profiles can be manipulated at the protein level in very different ways.

Analysis of transcription-level data is ultimately limited due to the lack of information about post-transcriptional or post-translational regulation. However, integration of transcriptome signatures with other sources of genomic and proteomic evidence (e.g. protein-protein interactions or phenotype data) can address this limitation. To search for network-level mechanisms explaining the observed changes in expression, genes exhibiting stem-cell associated changes in expression were overlayed on a mouse functional genetic network consisting of over 20000 linkages among protein-coding genes based on a diverse collection of genomic and proteomic data (Guan et al. PLoS Comput Biol. 2008). Integration of transcriptome data with this comprehensive functional network was used to corroborate the results from ARACNE.

These novel associations predicted by ARACNE and the functional network can be used to develop new hypotheses to test experimentally. This demonstrates that the vast amount of diverse public transcriptome data can be utilized to gain novel insights into stem cell-associated complex biological processes through the use of statistical and functional genomics tools.

See more of this Session: Genomic Approaches to Systems Biology
See more of this Group/Topical: Topical A: Systems Biology

Identifying Tightly Regulated and Variably Expressed Pathways by Differential Rank Conservation (DIRAC)

James A. Eddy, University of Illinois, Urbana, IL

Donald Geman, Johns Hopkins University, Baltimore, MD

Nathan D. Price, University of Illinois, Urbana, IL

Methods for analyzing high-throughput microarray data are shifting towards an increased focus on biologically meaningful pathways. Pathway-based gene expression analysis typically focuses on identifying related sets of genes that are individually or collectively over (or under) expressed in different phenotypes. Differential Rank Conservation (DIRAC) is a novel approach for studying gene ordering within pathways and is based on the relative expression ranks of participating genes. DIRAC provides quantitative measures of how pathway rankings differ both within and between phenotypes. DIRAC between pathways in a selected phenotype contrasts the scenarios where either (i) pathways are ranked similarly in all samples (high rank conservation); or (ii) the ordering of pathway genes is highly varied (low conservation). We examined a number of disease phenotypes including cancer subtypes and neurological disorders and identified pathways that appear to be tightly regulated based on high conservation of gene ordering. Tightness of regulation for a selected pathway can best be seen as the amount of variation in gene expression levels that is allowed for by the cell. Whereas studying up- or down-regulation of a pathway provides a specific measure of how a process may be modified, identifying pathways that are tightly or loosely regulated indicates the level of control across samples in a population. Pathways under tight control in a phenotype may be important for maintaining a specific cellular function and would therefore be important targets of study in different diseases. Interestingly, we observed a strong trend of greater shuffling overall (loose regulation of ordering across many pathways) in more malignant phenotypes and later stages of disease. This global pattern of increased disorder with malignancy highlights the utility of studying gene ordering within pathways, and also reveals a striking phenomenon that will drive future investigation and may lead to new understandings of gene expression in disease. The second form of DIRAC manifests as a change in ranking (i.e., shuffling) between phenotypes for a selected pathway. Variably expressed pathways identified by DIRAC serve as signatures for molecular classification, and were used to identify expression profiles with accuracies above 95% in many cases. The ability to accurately classify microarray samples provided strong validation for the pathway-level expression differences identified by DIRAC. The DIRAC approach for identifying expression profiles is noteworthy because it (i) is independent of data normalization; (ii) results in an elegant classifier where binary phenotype (e.g. disease) diagnosis can be done based simply on whether the score is computed to be above or below zero (and is bounded by 1 and -1); and (iii) appears thus far to be comparably accurate to state-of-the-art classification methods. By learning how the ranking of genes within pathways are regulated in different states, we can identify molecular signatures to distinguish between phenotypes and to aid in the identification of potential therapeutic targets.

See more of this Session: Genomic Approaches to Systems Biology
See more of this Group/Topical: Topical A: Systems Biology

Development of a Constraint-Based Model for Photobiological Production of Hydrogen in *Cyanothece* Sp. ATCC 51142

Trang T. Vu, Department of Chemical and Biological Engineering, University of Wisconsin-Madison, Madison, WI

Sergey Stolyar, Biological Sciences Division, Pacific Northwest National Laboratory, Richland, WA

Alex Belieav, Biological Sciences Division, Pacific Northwest National Laboratory, Richland, WA

Andrei Osterman, Burnham Institute for Medical Research, La Jolla, CA

Jessica De Ingeniis, Burnham Institute for Medical Research, La Jolla, CA

Jennifer L. Reed, Department of Chemical and Biological Engineering, University of Wisconsin-Madison, Madison, WI

Energy crisis is one of the world's major problems. For many years, energy production has heavily relied on the availability of fossil-fuels which are becoming increasingly limited and expensive. Research investments and efforts have been extensively focused on the development of clean and efficient processes for sustainable energy sources, of which hydrogen is considered as one of the promising future fuels (1). Among bio-based hydrogen-producing processes, oxygenic photosynthesis (where oxygen is produced as a by-product) receives great interest for it makes use of the abundance of sunlight and carbon dioxide, and thus produces hydrogen essentially at no substrate cost. However, oxygen produced in this process irreversibly inhibits nitrogenase and hydrogenase enzymes that are responsible for hydrogen production. Alternatively, indirect photosynthesis in which oxygen and hydrogen production are temporally or spatially separated can resolve the enzyme inhibition issue. This process however is poorly understood and the reported efficiency of hydrogen production is rather low (0.355mmol/L-hr in *Anabaena variabilis*) (2). Therefore, developing a model for this process is critical in not only understanding the physiology of the process but also identifying factors that impact hydrogen production rates. Constraint-based stoichiometric models (3) have been useful in biological discovery (4) and metabolic engineering to identify strains with significantly improved phenotypes (5-7).

Here we will present the reconstruction of a genome-scale metabolic network for *Cyanothece* sp. ATCC 51142, a cyanobacterium that can temporally separate the evolution of oxygen and hydrogen upon light-dark growth conditions. We used a RAST server (8) to generate, and integrate in the SEED database (9), an initial set of subsystems-based gene annotations for *Cyanothece* sp. that were further manually refined and improved while constructing the metabolic model. This model currently includes 798 genes, 682 proteins, 630 metabolites and 656 reactions accounting for common pathways such as central metabolism, nucleotide and amino acid biosynthesis, and those that are more unique to cyanobacteria such as photosynthesis, carbon fixation and cyanophycin production. The resulting model can produce hydrogen, and utilize ammonia, nitrate and molecular nitrogen as different nitrogen sources for *in silico* growth. Photosynthesis was modeled as sequential reactions that occur in each photosystem, I and II, in order to study the effect of different light wavelengths, and separate photosystem activities on cellular growth and hydrogen production rate.

Fixing metabolic gaps is an important step in reconstructing metabolic networks, especially if the gaps prevent the production of metabolites which are essential for cellular growth. For example, *Cyanothece* sp. ATCC 51142 lacks the gene encoding L-threonine deaminase, which catalyzes the conversion of L-threonine to 2-oxobutanoate, a precursor of isoleucine biosynthesis. Subsequently, our initial reconstruction was unable to synthesize isoleucine and this metabolic gap was fixed by including an alternative pathway for 2-

oxobutanoate biosynthesis from pyruvate and acetyl-CoA, the pathway similar to that found in *Geobacter sulfurreducens* (10). We are currently using various bioinformatic and comparative genomic techniques, such as the analysis of conserved operons and regulons, to identify candidate genes filling-in other gaps in our metabolic reconstruction. In addition to gap-filling, such analysis of genome context often allows for the prediction of new metabolic genes and pathways contributing to the further refinement and expansion and of the metabolic model.

It has been shown that with the accumulation of glycogen in the light phase, there is a burst of glycogen degradation in the dark phase that results in an increasing level of nitrogen fixation and respiration (11). The reconstructed metabolic network and its corresponding constraint-based model can be used to study the correlation between glycogen accumulation and degradation with hydrogen production. The predictions from the constraint-based model will be useful for improving hydrogen production using metabolic engineering approaches.

References 1. Dunn S. Hydrogen futures: Toward a sustainable energy system. *International Journal of Hydrogen Energy* **27**, 235-264 (2002).

2. Levina DB, Pitt L, & Love M. Biohydrogen production: Prospects and limitations to practical application. *International Journal of Hydrogen Energy* **29**, 173-185 (2004).

3. Palsson BO. *Systems Biology: Properties of Reconstructed Networks.* Cambridge University Press, New York, 12-25 (2006).

4. Reed JL, *et al.* Systems approach to refining genome annotation. *Proc Natl Acad Sci* U S A **103**, 17480-17484 (2006).

5. Alper H, Jin YS, Moxley JF & Stephanopoulos G. Identifying gene targets for the metabolic engineering of lycopene biosynthesis in *Escherichia coli. Metab Eng* **7**, 155-164 (2005).

6. Park JH, Lee KH, Kim TY & Lee SY. Metabolic engineering of *Escherichia coli* for the production of L-valine based on transcriptome analysis and *in silico* gene knockout simulation. *Proc Natl Acad Sci* U S A **104**, 7797-7802 (2007).

7. Fong SS, *et al. In silico* design and adaptive evolution of *Escherichia coli* for production of lactic acid. *Biotechnol Bioeng* **91**, 643-648 (2005).

8. Aziz RK, *et al.* The RAST server: Rapid annotations using subsystems technology. *BMC Genomics* **9**, 75 (2008).

9. Overbeek R, *et al.* The subsystems approach to genome annotation and its use in the project to annotate 1000 genomes. *Nucleic Acids Res* **33**, 5691-5702 (2005).

10. Risso C, Van Dien SJ, Orloff A, Lovley DR & Coppi MV. Elucidation of an alternate isoleucine biosynthesis pathway in *Geobacter sulfurreducens. J Bacteriol* **190**, 2266-2274 (2008).

11. Welsh EA, *et al.* The genome of *Cyanothece* 51142, a unicellular diazotrophic cyanobacterium important in the marine nitrogen cycle. *Proc Natl Acad Sci* U S A **105**, 15094-15099 (2008).

See more of this Session: Genomic Approaches to Systems Biology
See more of this Group/Topical: Topical A: Systems Biology

Convergent Transcription and Transcriptional Interference as a Regulatory Element in the Rhamnose Regulon of Escherichia Coli

Lon Chubiz, Chemical and Biomolecular Engineering, University of Illinois, Urbana, IL

Supreet Saini, Chemical and Biomolecular Engineering, University of Illinois, Urbana, IL

Christopher V. Rao, Department of Chemical Engineering, University of Illinois, Urbana, IL

The behavior of a circuit may also be determined by how it is physically encoded within the genome. By physical encoding, we mean the orientation and relative proximity of a transcription factor to its target structural genes. In particular, a given regulatory mechanism involving multiple genes can be physically arranged in any number of ways. The specific location of the genes may, therefore, represent an additional factor determining circuit behavior. In other words, two circuits involving the same genes and regulatory mechanism can potentially behave differently if they are physically encoded in alternate configurations.

In this work, we investigated the role of transcriptional interference due to convergent promoters using the rhamnose (rha) system from Escherichia coli as our model. RhaR and RhaS are transcriptional activators that respond to availability of L-rhamnose in the surrounding media. RhaR activates rhaSR operon and RhaS activates transcription of operon that encodes enzymes essential for L-rhamnose catabolic enzymes, rhaBAD, and the operon that encodes the L-rhamnose transporter, rhaT. The genes in this system are arranged in such a way that rhaSR and rhaT are expressed from convergent promoters with the C-terminals of rhaR and rhaT overlapping.

We report the effect of transcriptional interference between rhaR and rhaT on the dynamics of gene expression in the rhamnose utilization network in E. coli. Using a number of genetic constructs, we eliminated transcriptional interference between rhaR and rhaT and demonstrate that this leads to a faster and greater induction of the rhaBAD operon. Based on our results, we conclude that the physical arrangement of the genes on the chromosome plays an important role in governing gene expression dynamics.

See more of this Session: Genomic Approaches to Systems Biology
See more of this Group/Topical: Topical A: Systems Biology

Understanding the Role of Autophagy in Filamentous Fungi

Mark R. Marten, Univ. of Maryland, Baltimore County, Baltimore, MD

Liming Zhao, Univ. of Maryland, Baltimore County, Baltimore, MD

Judith K. Pollack, Univ. of Maryland, Baltimore County, Baltimore, MD

Yonghyun Kim, Univ. of Maryland, Baltimore County, Baltimore, MD

Bill J. Moss, Univ. of Maryland, Baltimore County, Baltimore, MD

Usha Sripathineni, Univ. of Maryland, Baltimore County, Baltimore, MD

Steven D. Harris, U. Nebraska, Lincoln, Lincoln, NE

David Schaefer, Towson University, Towson, MD

The impact of filamentous fungi on human activity cannot be understated. Pathogenic fungi detrimentally impact both people and plants and are responsible for thousands of deaths and billions of dollars in crop losses annually. In contrast, beneficial species are used to produce billions of dollars of chemicals, foodstuffs, and pharmaceuticals annually. In all these situations, fungi are likely to experience some degree of starvation, where survival requires recycling internal components in a process called "autophagy." Autophagy has been extensively studied in a few model organisms, is ubiquitous in eukaryotic cells, and its components are strongly conserved from yeast to man. However, in spite of it's likely involvement in most fungal processes, relatively little information is available on autophagy, or its related effects, in filamentous fungi.

We have systematically deleted a number of putative autophagy genes from the model fungus *Aspergillus nidulans*, and are using a sophisticated set of experimental tools to assess subsequent phenotypes providing insight regarding gene function. Proteomic analysis shows that gene deletions at different points in the autophagy induction pathway affect a wide range of cellular processes. Atomic force and electron microscopy show autophagy leads to fungal cell walls which are stiffer, thinner and weaker, implying autophagic fungi may be more susceptible to fragmentation during large scale fermentation. Fungi lacking a particular autophagy gene (Anatg13) show a significant reduction in both growth and branching rates in complete minimal medium, implying autophagy may be active even in the absence of starvation. In addition, the Anatg13 deletion appears to impact morphology as these mutants show no significant change in branching rate in any of the various growth environments employed. This implies autophagy, or the AnAtg13 protein, plays a role in regulation of fungal branching.

See more of this Session: Proteomics & Metabolomic Approaches to Systems Biology
See more of this Group/Topical: Topical A: Systems Biology

A Metabolomics-Based Analysis of Central Carbon Metabolism in 14 Yeast Species

Mark P. Styczynski, School of Chemical and Biomolecular Engineering, Georgia Institute of Technology, Atlanta, GA

Dawn A. Thompson, Broad Institute of MIT and Harvard, Cambridge, MA

Jenna Pfiffner, Broad Institute of MIT and Harvard, Cambridge, MA

Courtney French, Broad Institute of MIT and Harvard, Cambridge, MA

Aviv Regev, Broad Institute of MIT and Harvard, Cambridge, MA

Metabolic processes within a cell produce the energy and small molecule building blocks necessary for all cellular functions, and central carbon metabolism - the key metabolic module governing energy sources - is the cornerstone of these processes. Despite this key role, we understand surprisingly little about the function and evolution of central carbon metabolism on a systems scale. In this project we perform a comparative experimental and computational study of the function, regulation, and evolution of central carbon metabolism in fourteen *Ascomycota* fungi spanning 300 million years of evolution. We systematically measure intracellular metabolite concentrations under glucose depletion and other environmental perturbations. We integrate these metabolite profiles with expression profiles collected in our lab under the same conditions for a multi-scale analysis of how these metabolic responses have evolved across the phylogeny. We identify functional metabolic modules of co-varying metabolites and genes in each species, and we use a novel computational approach to identify orthologous modules across species and reconstruct their evolution. This reconstruction predicts conserved functional entities and major evolutionary changes in metabolic function and regulation. This innovative comparative metabolomic approach provides insights into the function and evolution of distinct metabolic strategies and establishes a new systems-level approach for the study of metabolic systems.

See more of this Session: Proteomics & Metabolomic Approaches to Systems Biology
See more of this Group/Topical: Topical A: Systems Biology

High-Throughput Reconstruction and Optimization of 130 New Genome-Scale Metabolic Models

Christopher Henry, Mathematics and Computer Science, Argonne National Laboratory, Argonne, IL

Matt DeJongh, Computer Science Department, Hope College, Holland, MI

Aaron Best, Biology Department, Hope College, Holland, MI

Rick Stevens, Mathematics and Computer Science, Argonne National Laboratory, Argonne, IL

Genome-scale metabolic models have emerged as a crucial resource for translating detailed knowledge of thousands of distinct enzymatic processes into global predictions of organism behavior [1]. These models provide a means of predicting essential genes, organism phenotypes, gene expression patterns, organism response to mutation, and metabolic engineering strategies [2]. Despite the great demand for genome-scale metabolic models, the rate of model development has lagged far behind the rate of genome sequencing. To improve the rate of development of new models, we have integrated a biochemical database, a set of thermodynamic estimations [3], a gap filling algorithm [4], and a model optimization algorithm[5] with the SEED framework for updating, correcting, and propagating annotations across hundreds of genomes simultaneously [6]. Within this framework we have implemented an automated pipeline for the high-throughput reconstruction and optimization of genome-scale metabolic models of prokaryotes [7], and we have applied this pipeline to produce functioning models for a diverse set of 130 organisms across 14 bacterial subdivisions. On average, these models comprise of 957 reactions associated with 678 genes covering 21% of the organism genome. Application of the gap filling algorithm resulted in the addition of an average of 53 reactions with no known corresponding genes. Whenever gene essentiality or phenotyping data was available, the model optimization algorithm was applied producing models with prediction accuracies that exceed 90%. Analysis of the reactions added by the gap filling process resulted in the following key discoveries; (i) by identifying the biomass components causing a reaction to be added by the gap filling, we were able to refine our biomass definitions for every organism modeled; (ii) by identifying reactions added to many different models by the gap filling algorithm, we were able to determine portions of the metabolism of various genera for which additional annotation, curation, and experimental work is required; and (iii) by applying the completed models to the prediction of essential gene sets, we were able to identify the metabolic functions that were consistently essential in every organism as well as those functions that were essential in only a small subset of organisms.

References:

1. Feist AM, Herrgard MJ, Thiele I, Reed JL, Palsson BØ: Reconstruction of Biochemical Networks in Microbial Organisms. Nat Rev Microbiol 2009, 7(2):129-143.

2. Feist AM, Palsson BO: The growing scope of applications of genome-scale metabolic reconstructions using Escherichia coli. Nat Biotechnol 2008, 26(6):659-667.

3. Jankowski MD, Henry CS, Broadbelt LJ, Hatzimanikatis V: Group contribution method for thermodynamic analysis of complex metabolic networks. Biophys J 2008, 95(3):1487-1499.

4. Satish Kumar V, Dasika MS, Maranas CD: Optimization based automated curation of metabolic reconstructions. BMC Bioinformatics 2007, 8:212.

5. Henry CS, Zinner J, Cohoon M, Stevens R: iBsu1103: an improved genome scale metabolic model of B. subtilis based on SEED annotations. Genome Biol 2009:submitted.

6. Overbeek R, Disz T, Stevens R: The SEED: A peer-to-peer environment for genome annotation. Communications of the Acm 2004, 47(11):46-51.

7. DeJongh M, Formsma K, Boillot P, Gould J, Rycenga M, Best A: Toward the automated generation of genome-scale metabolic networks in the SEED. BMC Bioinformatics 2007, 8:-.

See more of this Session: Proteomics & Metabolomic Approaches to Systems Biology
See more of this Group/Topical: Topical A: Systems Biology

A Novel Outer Approximation Algorithm for the Global Optimization of Metabolic Networks

Carlos Pozo, Chemical Engineering, Universitat Rovira i Virgili, Tarragona, Spain

Gonzalo Guillén-Gosálbez, Chemical Engineering, Universitat Rovira i Virgili, Tarragona, Spain

Laureano Jimenez, Chemical Engineering Department, University Rovira i Virgili, Tarragona, Spain

Albert Sorribas, Basic Medical Sciences Department, Unversity of Lleida, Lleida, Spain

Abstract

The study of complex biological systems requires the integration of experimental and computational research by adopting a systems biology approach. Here, computational biology plays a major role by developing modeling tools that aim to provide a powerful foundation from which to address critical scientific questions. In this general context, we distinguish between different applications for which specific tools able to deal with the associated complexity are required. One of these applications, the optimization of metabolic networks, has recently emerged as a very important goal in biotechnology [2-3, 5-6, 9].

In recent years, the use of genetic manipulation techniques has led to significant improvements in the production of certain biochemical products. However, in most cases mutation and selection of new processes has been made in a proof-and-error basis [10], which can lead, at the best, to local optimal solutions. As actual biological processes are operating far from their global (theoretical) optimum, one expects that they could be further improved if quantitative design principles for enzyme activities were provided by a more rational approach like optimization. This optimization is known as *engineering design* and consists of, given a model, finding the appropriate changes in the gene expressions so that a certain objective function (typically a flow) is optimized.

One of the key steps in this approach is the selection of the appropriate mathematical model among the different representations available. Among them, representations that result from the combination of linear stoichiometric descriptions of the network and nonlinear empirical equations expressing the velocity show a good compromise between accuracy and simplicity. Particularly, in this group we find the S-System and General Mass Action (GMA) representations, which have been shown to capture in an efficient way the non-linearities associated with the regulatory processes of the networks while showing linear properties in the logarithmic space [10]. GMA models only differ from S-System models in the way in which the branching points are handled. In GMA models, each process in a branching point is approximated separately, while in S-System models, all the input and output flows are collected and modeled as just two contributions.

S-Systems models based on the power-law formalism were first used in metabolic optimization problems by Voit [11]. In S-Systems models, a logarithmic transformation is enough to obtain a linear representation providing a convenient framework for linear optimization techniques [1, 7, 11]. This is not possible with GMA models. The complexity in solving these models stems from their non-convexities, which give rise to multiple local optima in which standard optimization packages may get trapped during the search. Hence, advanced global optimization methods are required to converge to the global optimum of the problem. These methods are based on calculating valid lower and upper bounds on the global solution of the problem. These bounds are gradually tightened until a desired optimality criterion is satisfied.

Unfortunately, there are currently very few methods available for the global optimization of GMA models [8, 10]. The main drawback of these existing methods is that they provide solutions with large optimality gaps. Thus, there is still room for improvement to achieve the performance required by this type of studies.

Specifically, in this work we present a novel algorithm to globally optimize metabolic networks that are modeled via the GMA representation. The optimization method we propose, which is based on the works by

Polisetty et al. [10] and Bergamini and coworkers [4], makes use of an outer-approximation strategy. In the approach presented, the original problem is decomposed into two problems at two different hierarchical levels: a master problem and a slave problem. The master problem takes the form of a convex mixed-integer linear programming (MILP problem) and is obtained by convexifying the original non-convex NLP formulation. This MILP problem is a relaxation (i.e., it overestimates the original search space) of the original non-convex problem, and hence provides a valid upper bound on its global solution. The output of this master problem is used to set bounds for some of the variables in the non-convex slave problem, which is locally optimized to obtain a lower bound on its global optimum. The relaxation of the master problem is then gradually tightened and the whole procedure is repeated until a termination criterion is satisfied.

The capabilities of our strategy have been illustrated through its application to the optimization of (1) the ethanol production in the fermentation of *Saccharomyces Cerevisiae* and (2) the citric acid synthesis rate produced by *Aspergillus niger*. GMA models for these two systems can be found in [10]. The method presented provided solutions with much lower optimality gaps than those reported in the literature. Furthermore, our strategy was shown from these numerical examples to perform much better than BARON, the *state of the art* solver for global optimization, in terms of optimality gap and computation time.

References

[1] Alvarez-Vasquez F, Canovas M, Iborra JL, Torres NV: Modeling, optimization and experimental assessment of continuous L-(-)-carnitine production by Escherichia coli cultures. Biotechnol Bioeng 2002, 80(7):794-805.

[2] Bailey J, Birnbaum S, Galazzo J, Khosla C, Shanks J: Strategies and challenges in metabolic engineering. Ann. NY Acad. Sci. 1990, 589:1-15.

[3] Banga JR: Optimization in computational systems biology. BMC Syst Biol 2008, 2:47.

[4] Bergamini ML, Aguirre P, Grossmann I: Logic-based outer approximation for globally optimal synthesis of process networks. Computers and Chemical Engineering 2005, 29:1914-1933.

[5] Cameron D, Chaplen F: Developments in metabolic engineering. Curr. Opin. Biotechnol. 1997, 8:175-180.

[6] Cameron D, Tong J: Cellular and metabolic engineering: an overview. Appl.Biochem.Biotechnol. 1993, 38:105-140.

[7] Marin-Sanguino A, Torres NV: Optimization of biochemical systems by linear programming and general mass action model representations. Math Biosci 2003, 184(2):187-200.

[8] Marin-Sanguino A, Voit EO, Gonzalez-Alcon C, Torres NV: Optimization of biotechnological systems through geometric programming. Theor Biol Med Model 2007, 4:38.

[9] Mendes P, Kell D: Making cells work - metabolic engineering for everyone. Trends Biotechnol. 1996, 15:6-7.

[10] Polisetty PK, Gatzke EP, Voit EO: Yield optimization of regulated metabolic systems using deterministic branch-and-reduce methods. Biot Bioeng 2008, 99(5):1154-69.

[11] Voit EO: Optimization in integrated biochemical systems. Biotechnol Bioeng 1992, 40(5):572-82.

See more of this Session: Proteomics & Metabolomic Approaches to Systems Biology
See more of this Group/Topical: Topical A: Systems Biology

ciFBA: A Scalable Framework for Integrated Analysis of Metabolism and Regulation

Paul A. Jensen, Department of Biomedical Engineering, University of Virginia, Charlottesville, VA

Jason A. Papin, Department of Biomedical Engineering, University of Virginia, Charlottesville, VA

High-throughput data from genome sequencing and expression profiling have enabled the reconstruction of cellular regulatory and metabolic networks. While flux balance analysis and other metabolic reconstruction techniques have successfully predicted growth rates for several organisms using constraint-based linear programming, these methods lack an efficient means for seamlessly integrating a regulatory network to examine the full effect of a metabolic environment on an organism. This presentation will detail a novel, scalable extension to flux balance analysis, ciFBA (concurrent, integrated Flux Balance Analysis), that incorporates both a stoichiometric metabolic reconstruction and a constraint-based representation of Boolean gene regulatory network. ciFBA uses a single mixed integer linear program to optimize both an organism's internal flux distribution and external media composition; this approach is uniquely suited to efficiently calculate: (1) optimal and minimal growth media, (2) necessary environments for expression of a specific set of genes, and (3) media-specific growth rates. Many of these predictions can be verified by simple growth experiments, requiring only different media compositions. Ultimately, ciFBA's integration of regulation and metabolism will constitute a more complete description of an environment's effect on an organism, thereby creating a more flexible platform for *in silico* metabolic engineering.

See more of this Session: Proteomics & Metabolomic Approaches to Systems Biology
See more of this Group/Topical: Topical A: Systems Biology

Topological Metabolic Analysis for the Interrogation of Metabolic Networks

Adam C. Baughman, Chemical & Biological Engineering, Rensselaer Polytechnic Institute, Troy, NY

Lealon L. Martin, Chemical and Biological Engineering, Rensselaer Polytechnic Institute, Troy, NY

Susan T. Sharfstein, Department of Chemical and Biological Engineering, Rensselaer Polytechnic Institute, Troy, NY

We introduce Topological Metabolic Analysis (TMA) a novel, flexible, optimization-based framework for computational modeling of metabolic networks. The framework is adapted from the Infinite-Dimensional-State-Space approach originally introduced by Manousiouthakis and co-workers for the representation of complex networks of heat- and mass-exchangers. We claim that our approach presents certain mathematical advantages as compared to the stoichiometric balancing techniques in common use, and that these advantages are primarily conferred through a rigorous mathematical treatment of network topology. We demonstrate the ability of our technique to probe network reconstructions for potentially missing functionality. We further demonstrate the technique's unique ability not only to fully describe the topology of a particular metabolic network, but to identify a finite number of alternate network topologies that are all each globally optimal with respect to a given objective. Because this modeling framework implicitly enforces the metabolite balances typically elucidated using costly C-13/NMR experiments, it presents an \emph{in silico} means of calculating the complete set of reaction activities within a network without the need for costly and laborious in vitro experimentation. Finally, we demonstrate the ability of our modeling approach to identify network solutions which simultaneously optimize a particular teleological objective while ensuring minimal deviation from empirically observed system behavior through the use of a novel ``hybrid" objective function.

See more of this Session: Proteomics & Metabolomic Approaches to Systems Biology
See more of this Group/Topical: Topical A: Systems Biology

Qualitative Multiscale Methodology for Biological Networks Driven by Experimental and Clinical Data: Insulin Network in Beta Cells and Association with Diabetes

Kosta Makrodimitris, MAKROnanoKosmos, Beltsville, MD

Tom Woolf, Johns Hopkins University, Baltimore, MD

There is a significant increase in genomics, metabolomics and proteomics datasets and a growing demand to provide qualitative mathematical models for fighting obesity, diabetes and other diseases related to the metabolism and regulation. Qualitative approaches are efficient and adequate in many cases where there is limited quantitative information on reaction mechanisms for biological networks.

We have designed a multiscale methodology using network ideas (molecules-cells-tissues-body) to model biological networks in different length and time scales extending the simple boolean networks and avoiding complex stoichiometric approaches. Our method allows the creation and customization of networks based on limited user data and available bioinformatics databases. The discrete time steps, molecular states, and relative concentrations are described like a cellular automata model where each node(the molecule state and properties) is updated from the input and signals of the neighboring connected ones. The connections between the molecular nodes can be represented in an experimentally realistic way using linear or nonlinear connections allowing the description of complex mechanisms such as feedback loops.

We focus on the role of insulin which is made and released from the beta cells and controls the level of glucose in the blood. Insulin is associated with both types of diabetes and there are currently a lot of available data and a need to explain the complex interconnected mechanisms causing the disease. Simulations of signaling and metabolic pathways of insulin in beta cells and pancreas tissue will be presented and a comparison with experimental data and other approaches will be discussed.

BIBLIOGRAPHY: System Modeling in Cellular Biology: From Concepts to Nuts and Bolts, Edited by Zoltan Szallasi, Jörg Stelling and Vipul Periwal, 2006

See more of this Session: Proteomics & Metabolomic Approaches to Systems Biology
See more of this Group/Topical: Topical A: Systems Biology

AUTHOR INDEX

AUTHOR INDEX